Palgrave Studies in Classical Liberalism

This series offers a forum to writers concerned that the central presuppositions of the liberal tradition have been severely corroded, neglected, or misappropriated by overly rationalistic and constructivist approaches.

The hardest-won achievement of the liberal tradition has been the wrestling of epistemic independence from overwhelming concentrations of power, monopolies and capricious zealotries. The very precondition of knowledge is the exploitation of the epistemic virtues accorded by society's situated and distributed manifold of spontaneous orders, the DNA of the modern civil condition.

With the confluence of interest in situated and distributed liberalism emanating from the Scottish tradition, Austrian and behavioral economics, non-Cartesian philosophy and moral psychology, the editors are soliciting proposals that speak to this multidisciplinary constituency. Sole or joint authorship submissions are welcome as are edited collections, broadly theoretical or topical in nature.

Jonathan H. Adler
Editor

Climate Liberalism

Perspectives on Liberty, Property and Pollution

palgrave
macmillan

Editor
Jonathan H. Adler
School of Law
Case Western Reserve University
Cleveland, OH, USA

ISSN 2662-6470 ISSN 2662-6489 (electronic)
Palgrave Studies in Classical Liberalism
ISBN 978-3-031-21107-2 ISBN 978-3-031-21108-9 (eBook)
https://doi.org/10.1007/978-3-031-21108-9

This Palgrave Macmillan imprint is published by the registered company Springer Nature Switzerland AG
The registered company address is: Gewerbestrasse 11, 6330 Cham, Switzerland

Acknowledgments

This book grew out of a collaborative project between the Institute for Humane Studies and the Coleman P. Burke Center for Environmental Law at Case Western Reserve University School of Law exploring what, if anything, classical liberalism had to offer in the search for effective and meaningful climate change policy. Among other things, the project included a discussion colloquium and interdisciplinary workshop. Funding for this project was provided by the John Templeton Foundation and the Institute for Humane Studies.

Many people contributed to making the aforementioned programs successful and helping to lay the foundations for this volume. They include Michael Brodrick, Eric Claeys, Allison Grant, Patricia Harbold, Dagney Hatfield, Andrew Morriss, Maria Rogacheva, Lyman Stone, Katherine Wright, Marty Zupan, and especially Greg Wolcott, whose contributions were essential at all stages of this work. Thanks are also due to Leslie Marsh, who believed this volume would make a useful contribution to the Palgrave Studies in Classical Liberalism series, and those at Palgrave and Springer who helped shepherd this project to completion, including Ellie Duncan and Hemapriya Eswanth.

In addition to the above, I would like to thank Akua Oppong and Kory Roth for research and administrative assistance in getting the manuscript together, Elissa Tenant for administrative support, Lisa Peters in the Case Western Reserve University Law Library, and anyone else who I am neglecting who helped me get this project across the finish line.

And, as with everything I do, this would not be possible without the continuing love and support of my family, for which I am eternally grateful.

Contents

Notes on Contributors

Jonathan H. Adler is the inaugural Johan Verheij Memorial Professor of Law and Director of the Coleman P. Burke Center for Environmental Law at the Case Western Reserve University School of Law, and a Senior Fellow at the Property and Environment Research Center. His previous books include *Marijuana Federalism: Uncle Sam and Mary Jane*, *Business and the Roberts Court*, and *Rebuilding the Ark: New Perspectives on Endangered Species Act Reform.*

Karen Bradshaw is Professor at the Sandra Day O'Connor College of Law and Senior Global Futures Scientist for the Julie Ann Wrigley Global Futures Laboratory at Arizona State University.

Mark Budolfson is currently Assistant Professor in Environmental Health Sciences, Population-Level Bioethics, and Philosophy at Rutgers University, and Associate Member of the Princeton University Climate Futures Initiative, as well as a Faculty Affiliate at the University of Vermont Gund Institute for Environment.

Billy Christmas is Lecturer in Political Theory at King's College London in the Department of Political Economy, where he is also the PPE Programme Director.

Daniel H. Cole is the Robert H. McKinney Professor of Law at Indiana University's Maurer School of Law. He is also an adjunct professor for Public and Environmental Affairs and the Department of Political Science.

David Dana is the Kirkland & Ellis Professor of Law and Director of the Program on Sustainability and Food and Animal Law at the Northwestern Pritzker School of Law. He is also a professor of Real Estate at the Northwestern Kellogg School of Management.

Ed Dolan is Senior Fellow at the Niskanen Center.

Monika Ehrman is Associate Professor of Law at the University of North Texas at Dallas College of Law and serves on the Board of Directors for the Foundation for Natural Resources and Energy Law and Matador Resources Company.

Brian T. Fitzpatrick is the Milton R. Underwood Chair in Free Enterprise and Professor of Law at Vanderbilt Law School.

Andrew P. Morriss is the Dean of Texas A&M University School of Innovation and Vice President for Entrepreneurship & Economic Development at Texas A&M University.

Mark Pennington is the Director of the Centre for the Study of Governance and Society and a professor of Political Economy and Public Policy in the department of Political Economy.

Dan C. Shahar is Visiting Scholar at Tulane University.

Catherine M. Sharkey is the Segal Family Professor of Regulatory Law and Policy at the New York University School of Law.

John Thrasher is Associate Professor in the Philosophy Department and the Smith Institute for Political Economy and Philosophy at Chapman University and an Adjunct Senior Research Fellow at Monash University.

List of Figures

List of Tables

Introduction

Jonathan H. Adler

The environmental consequences of modern civilization are profound. Increasing human prosperity has come at an ecological cost, including the disruption of ecosystems, accumulation of waste products, and emission of pollutants into water and air. Yet the same economic development that has contributed to such problems has also generated the wealth, technology, and desire to address environmental concerns. The last half century has seen substantial environmental progress as human lifespans have increased, poverty has declined, and Malthusian fears of resource depletion have been dispelled.[1] Nonetheless, significant environmental challenges remain, particularly as they relate to pollution.

Of those challenges, climate change is perhaps the most pressing. Carbon-based fuels have provided access to affordable energy throughout much of the world, fostering greater human prosperity. At the same

J. H. Adler (✉)
Case Western Reserve University School of Law, Cleveland, OH, USA
e-mail: edolan@niskanencenter.org

J. H. Adler (ed.), *Climate Liberalism*, Palgrave Studies in Classical Liberalism,
https://doi.org/10.1007/978-3-031-21108-9_1

time, the combustion of such fuels has contributed to a dramatic increase in atmospheric concentrations of greenhouse gases—concentrations that threaten to increase global average temperatures, raise sea levels, disrupt weather patterns, and alter agricultural productivity, among other things.[2] Yet because greenhouse gas emissions are so ubiquitous, and because energy is so important for human flourishing, climate change presents a particularly "wicked" problem. It "defies resolution because of the enormous interdependencies, uncertainties, circularities, and conflicting stakeholders implicated by any effort to develop a solution."[3]

Climate change also confounds the classical-liberal intellectual tradition, particularly its emphasis on property rights and decentralized authority. While classical-liberal thinkers have explained how competitive markets and property rights encourage efficiency, innovation, and sustainable utilization, less attention has been paid to the more difficult environmental challenges that have come to the fore in the twenty-first century.[4] In particular, classical-liberal thinkers have, to date, invested relatively little effort into examining how liberal principles might inform approaches to more difficult pollution problems in general, and climate change in particular.[5]

Climate Liberalism seeks to fill this gap by examining the extent to which classical-liberal principles, including an emphasis on property rights, decentralized authority, and dynamic markets, can inform policy approaches to large-scale pollution problems, including climate change. The contributors represent multiple academic disciplines and perspectives. Some would consider themselves to be classical liberals, while others would not. All believe that it is worthwhile to examine how such principles might inform the development of policies that would adequately address large-scale environmental problems while preserving individual liberty and maintaining a free and dynamic economic marketplace. Chapters consider the role of property rights and common-law legal systems in controlling pollution, the extent to which competitive markets backed by legal rules encourage risk minimization and adaptation, and how to identify the sorts of policy interventions that may help address climate change in ways that are consistent with liberal values. This introduction sets the stage with an overview of classical-liberal approaches to environmental protection, including where such approaches have succeeded and failed, and an explanation of why it is necessary to evaluate the potential for such approaches to inform pollution control policies in general and climate-change policy in particular.

Conventional Environmental Protection

The conventional approach to environmental problems has been to adopt administrative regulations to control the environmental consequences of productive activity. Government agencies own and manage threatened resources and regulate the use of others. Rules are placed upon those activities that modify landscapes, generate wastes, or result in the emission of pollutants, as well as on many activities that merely threaten such consequences. Prescriptive regulatory approaches have produced some gains, particularly in the area of air pollution control, but there are also reasons to suspect such approaches have their limits. As the low-hanging fruit has been picked, the environmental challenges that remain often defy easy administrative fixes.[6] In some cases, prescriptive environmental regulations even exacerbate the environmental problems they are supposed to solve by undermining incentives for responsible environmental stewardship or constraining the adoption of measures or technologies that could reduce ecological harms or risks to human health.[7]

As noted environmental law scholar Richard Stewart has observed, centralized environmental regulation is inherently limited by "the inability of central planners to gather and process the information needed to write directives appropriately responsive to the diverse and changing conditions of different economic actors, and the failure of central planning commands to provide the necessary incentives and flexibility for environmentally and economically beneficial innovation."[8] If governmental intervention is justified every time there is a "market failure," such as the failure to account for "externalities" resulting from productive activity, there are few limits on when and where the government may intervene.[9] Virtually all human activity has some environmental consequence, and yet it would be impossible for a regulatory agency to plan and account for all such effects. If economic central planning is destined for failure, ecological central planning does not have much hope.[10]

In addition to the practical limitations of the conventional approach to environmental protection, it is also anathema to many classical liberals who value economic freedom and individual liberty. The conventional approach is based on expansive governmental interventions in the economy and limits on personal behavior. The ecological mantra that "everything is connected to everything else" challenges the very idea that there are spheres of private autonomy that should lie beyond the reach of the state. But what is the alternative?

Classical-Liberal Environmental Protection

A classical-liberal environmentalism is one that focuses on how liberal institutions can advance environmental values. Classical-liberal approaches to environmental concerns are more likely to start with the premise that environmental problems are not the result of "market failure," but of a failure to have markets—or, more properly, a failure to extend and protect property rights and voluntary exchange to the full range of ecological resources.[11] The role of government is not to identify and address each and every environmental impact of productive activity, but to provide and preserve the underlying institutional framework that protects the persons and property of individuals while allowing people to pursue and advance environmental values.

Much twentieth-century environmental thought is suspicious of markets and economic growth, believing both pose threats to environmental protection. However, this suspicion may be unfounded. Markets create economic pressure for more efficient resource use, encourage technological and institutional innovations that reduce the adverse consequences of productive activity, and generate incentives for firms to satisfy the environmental preferences of individuals—preferences that appear to increase as people become more wealthy.[12] Competitive markets take advantage of the dispersed knowledge possessed by individuals about their own circumstances and subjective value preferences, as well as the availability of and demand for resources.[13] At the same time, property rights systems preserve a relatively large sphere of individual autonomy and reinforce notions of personal responsibility.

Liberal market-oriented societies are not only more prosperous than their alternatives, but they also appear to experience more positive environmental outcomes.[14] Both economic growth rates and key measures of environmental quality are greater in countries where property rights are protected.[15] Greater protection of property rights correlates with a decline in deforestation as well as access to safe drinking water and sanitation services.[16] More broadly, higher levels of economic freedom are "associated with better environmental and public health outcomes."[17]

The record of overcoming resource scarcity and encouraging more sustainable use of resources in liberal societies is nothing short of remarkable.[18] This is particularly true when compared to the record of illiberal societies.[19] Dematerialization may be the most important, yet unsung, example of environmental progress in the twenty-first century. A soda can today is made with a fraction of the metal required 50 years ago. More significantly, dynamic markets have driven dematerialization to the point where we are observing a fundamental decoupling of resource consumption from economic growth, such that as mature, market-oriented economies grow, they not only use fewer resources per unit of output, but they also consume fewer resources overall.[20] In short, economic growth in the most developed nations increasingly coincides with a net reduction in resource consumption. Yet such environmental successes are often ignored because there is no policymaker or program that can take credit for them. They are the result of market processes, not governmental direction or design.

Markets rely upon the existence of property rights, and the creation of property rights in ecological resources has encouraged more sustainable use and has helped overcome the "tragedy of the commons."[21] As a general rule, where resources are owned, they are more likely to be used in a sustainable fashion. Property owners have both the ability to protect the owned resource and a substantial incentive to ensure that the value of their property—both to themselves and to others—is maintained.[22] As Harold Demsetz explained, "If a single person owns land, he will attempt to maximize its present value by taking into account alternative future time streams of benefits and costs and selecting that one which he believes will maximize the present value of his privately-owned land rights."[23] Owners care about the value of what they own, both to themselves and to others, and both now and in the future. Conversely, the lack of property rights in the underlying resource leaves users with little incentive to engage in conservation efforts, as any such efforts are likely to be wasted.[24]

Consistent with standard economic theory, those resources that have been incorporated into market institutions through property rights are managed more sustainably than those managed by governments or left in open-access commons.[25] To take just a few examples, privately owned forests exhibit higher rates of forest growth and tree planting than those managed by government agencies,[26] and oyster beds that are privately managed tend to be healthier and more productive than their publicly managed counterparts.[27] Even quasi-property rights in wildlife, such as elephants in sub-Saharan Africa, have furthered conservation.[28] Such measures have also generated positive externalities as well, as efforts to protect wildlife in the wild have resulted in greater protection for broader ecosystems. This is not to say such arrangements are perfect. The point is simply that liberal institutions, where they can be implemented (an important qualification), typically represent a superior choice to available alternatives.[29]

Perhaps the clearest evidence for the value of creating property rights in ecological resources comes from marine fisheries.[30] For decades, marine fisheries were plagued by overfishing. Despite extensive regulatory efforts, many of the world's largest fisheries declined drastically, and others ultimately collapsed. Beginning in the late twentieth century, however, many countries began to experiment with property-based management regimes, often called "catch shares." A 2008 review of such programs around the world found that implementing such programs "halts, and even reverses, the global trend toward widespread [fisheries] collapse."[31] Such programs covered only two percent or so of fish stocks around the world, but as of 2010, they accounted for approximately twenty-five percent of the volume of fish caught annually worldwide.[32] Subsequent research has confirmed that the creation of property-based management systems encourages greater stewardship among fishery participants, including efforts to maintain and enforce sustainable limits on the total catch.[33] Property rights have made fisheries more sustainable, even as population growth and economic development have increased demand for fish.

The experience with fisheries suggests the value of learning how property rights may be extended to threatened ecological resources, particularly those we wish to simultaneously exploit and conserve. The experience also suggests how challenging such efforts can be. Only decades after fishery economists began to identify potential mechanisms for the extension of property rights to fisheries did such reforms begin to get adopted, and only recently—in the past two decades or so—has conclusive empirical evidence on the value of these approaches emerged. Following this model for other resources will take no less effort, and the vindication of classical-liberal ideas about how to protect other resources is by no means assured.

There have also been economic and ecological gains from the development of markets in water through the recognition of transferable property rights and the adoption of market pricing.[34] As a consequence of such reforms, the volume of water trades, leases, and purchases has been increasing in the United States.[35] More importantly, the development of water markets, where allowed, has encouraged greater water conservation and has facilitated the allocation of water for species protection and other environmental purposes. As various jurisdictions have begun to recognize rights in water for non-consumptive uses—such as instream flows used for conservation purposes—conservation organizations have been able to enter the market to purchase or lease water rights for the benefit of threatened fish populations.[36] The recognition of property rights in water gives farmers a potentially marketable asset, and the demand for instream rights from conservationists, recreationists, and others creates a financial incentive to "use" water in ways that benefit species and local ecosystems.[37] As a consequence, where robust water rights are recognized, voluntary, cooperative transactions to reallocate water have begun to replace lobbying and political maneuvering. Rather than seek the imposition of additional regulatory controls which may trigger conflict and litigation, conservation organizations can negotiate with farmers and ranchers to purchase, lease, or otherwise transfer water rights.

Markets in ecological resources can also facilitate adaptation to changing ecological conditions, as might occur due to climate change.[38] Indeed, the primary virtue of markets is not the generation of static efficiency, but the constant pressure to allocate resources to their highest valued uses, even as the value of competing uses changes over time. Markets are an immensely powerful means of discovering and aggregating time- and place-specific information, including subjective value preferences, and markets are constantly adapting as such information, or the conditions upon which it is based, evolves.[39] In the case of water, for instance, market prices communicate information about the relative supply and demand for particular uses in particular places. Price changes encourage resource owners to consider voluntary transactions to transfer resources to higher valued uses. Even the United Nations' Intergovernmental Panel on Climate Change has noted that "improving the functioning of water markets could help create the kind of flexibility needed to respond to uncertain changes in future water availability" caused by climate change.[40]

The Problem of Pollution

Property-based reforms have encouraged sustainable resource use and conservation. Nonetheless, they have not been as successful at controlling pollution. It is easier to protect property rights from expropriation or physical occupation than from airborne or waterborne intrusions or despoilation of unowned environmental commons.[41] Market pressures for more efficient resource use contributed to pollution declines in the late twentieth and early twenty-first centuries, but it is nonetheless difficult to attribute water and air quality improvements to the operation of property rights and markets alone.

Property rights, to be meaningful, must be protected against private harm, and no individual property owner has the right to use his or her property in a fashion that harms or interferes with the ability of his or her neighbor to do the same. When A's use of her property interferes with B's ability to make equivalent use of his property, B's property rights have been violated.[42] From a rights-based perspective, it is not the generation of waste or effluent that creates the violation, but the imposition of such waste onto the person or property of another. And, like other rights violations, pollution can justify government action to assist in the protection and defense of rights, whether by direct efforts to protect such rights and prosecute their violation, or prescriptive measures to prevent such violations in the first place.

As explained by Terry Anderson and Donald Leal, "The free market environmental approach to pollution is to establish property rights to the pollution disposal medium and allow owners of those rights to bargain over how the resource will be used."[43] In the American legal system, pollution can be addressed by principles of trespass and nuisance.[44] This approach may work well where the source of pollution is easily identified, the number of polluters or pollution victims remain small, the resulting harm is apparent, and there are adequate judicial or other dispute resolution systems in place. So if a factory is belching smoke or refuse streams from an uncontrolled pipe, it may be easy to show who is responsible for the pollution, to quantify the resulting harm, and seek legal redress. Yet if the pollution in question is barely perceptible, comes in the form of trace amounts of difficult to measure but particularly harmful contaminants, or only manifests harm over time, the burdens of proof in legal proceedings may prevent any meaningful redress. The same can be said if the ultimate source of the pollution is some distance away. In such cases, protecting property rights against pollution can be a challenge.[45] (Admittedly, such challenges may frustrate the effectiveness of administrative regulation as well.)

The complexity of many pollution problems confounds any effort to control pollution through the assertion and enforcement of property rights in land. The problem is that many, if not most, of the pollution problems of concern today are not so simple and straightforward as the factory example above. The demand for environmental regulation does not arise from bilateral disputes between landowners in which one is causing an emission or waste stream to intrude upon the person or property of another. Rather, regulations are developed to address instances in which there are many alleged polluters affecting many parcels of land—what may be conceived of as large-number externality problems—when the degree and nature of causation is difficult to monitor and observe, and when the risks posed by allegedly polluting actions may be latent or uncertain.

In all but the simplest scenarios, neither legal enforcement of property rights against pollution nor low transaction costs can be assumed. Property rights in many environmental media, most notably water and air, are incomplete and undefined. This complicates the development of Coasean contractual solutions to pollution problems. Traditional dispute resolution systems, such as common-law adjudication, may work tolerably well for relatively simple, bilateral pollution disputes where the relevant rights are well-defined and the transaction costs are relatively low, but they may falter when the pollution at issue is generated by many sources or affects many properties. In cases in which there are many polluters, many victims, or both, transaction costs escalate rapidly and traditional property rights solutions falter.

The difficulty of addressing large-scale or dispersed pollution problems through the definition and enforcement of property rights does not necessarily mean that the adoption of administrative regulation is superior. Centralized regulatory systems, even if adorned with "market mechanisms," face limitations of their own, and government interventions may be compromised or misdirected by interest group pressures.[46] An imperfect policy approach may still be superior to the alternative, and there are many reasons to be suspicious of centralized environmental regulation. Nonetheless, it is fair to note that addressing pollution problems may require the consideration of policy measures that classical liberals might otherwise abjure.[47]

The Climate Challenge

As difficult as many conventional pollution problems may be for classical-liberal principles, the challenge posed by climate change is even greater. The global atmosphere is, in many respects, the mother of all commons problems, and there is no clear way to apply classical-liberal principles to the threat of climate change. Failure to do so, however, presents its own threat to classical-liberal values, as both climate change itself and conventional climate-change policies pose threats to principles of individual liberty and limited government.

The leading greenhouse gas, carbon dioxide, is a ubiquitous by-product of modern civilization. Measures to control it necessarily implicate energy use throughout the economy. While scientists may debate the precise degree to which human activity is contributing to a gradual warming of the atmosphere, there is little dispute that human activity has contributed to such warming and will continue to do so in the future. Even warming "skeptics" admit there is an anthropogenic contribution to warming and that such warming will contribute to phenomena, such as sea-level rise, that present serious concerns.[48] Yet there is no obvious or easy way to assign property rights in the atmosphere, no ready mechanism to facilitate meaningful Coasean bargaining, and only limited opportunities to seek legal redress for warming-induced harms.

It is important to stress that even modest increases in global temperature can be expected to produce effects, such as sea-level rise, which readily justify government action under classical-liberal principles.[49] It is a well-established principle in the Anglo-American legal tradition that one does not have the right to use one's own property in a manner that causes harm to one's neighbor. There are common-law cases going back 400 years establishing this principle, and international law has long embraced a similar norm. Further, those most responsible for the accumulation of greenhouse gases in the atmosphere to date—people in wealthy, developed nations—are not those most vulnerable to the most likely and most severe effects of climate change, sea-level rise in particular.[50]

Some classical liberals may object on the grounds that the costs of any government interventions to mitigate the threat of climate change will outweigh any potential benefits, yet such objections are difficult to square with any principled commitment to property rights.[51] Even if one believes that suffering from the potential effects of climate change should be weighed against losing the benefits of increased prosperity by actively intervening in the economy to reduce emissions, that would not justify running roughshod over the rights of those adversely affected by warming-induced flooding and other consequences of climate change. In other contexts, such as debates over eminent domain, classical liberals recognize the danger of allowing promises of economic prosperity to trump individual property rights. There is no reason to apply a different principle in the climate context, particularly when (unlike in the case of eminent domain), those whose property rights are violated receive no direct compensation. Property rights should not be sacrificed as part of some anticipatory utilitarian calculus.

Further, even if one believes the likelihood of catastrophic climate change is small, the consequences are sufficiently grave to justify prudent measures to reduce the likelihood and magnitude of adverse events. The reason to install smoke alarms, acquire fire extinguishers, and buy home insurance is not that you expect a house fire. Rather, such steps represent prudent investments to protect against low-probability/high-magnitude events. The costs and dislocation resulting from a catastrophic house fire are sufficiently grave that they justify prudent investments that will reduce the likelihood and the consequences of such an event.

A similar logic applies to many larger risks that require collective action, as classical liberals readily recognize in other contexts. Government spending on national defense, for instance, is not justified because conflict with another country is certain. Rather, such investments (and other national security measures) are justified as steps to make international conflict less likely, and the consequences of any such conflict less severe. But such justifications for government action also have limits. Acknowledging the need for government involvement does not mean giving the state a blank check.

Climate change presents a similar concern. While future generations may be sufficiently wealthy and technologically advanced to deal with many consequences of climate change, there is a sufficient probability of large-scale risks to justify some mitigation measures. This is particularly so because climate change represents a non-diversifiable risk, in that the worst climate outcomes will also compromise the economic and technological ability to respond to them.[52] The question then becomes what sorts of measures can be justified and whether some forms of government intervention are more consistent with classical-liberal principles than others.

Nonetheless, the threat posed by climate change, and problems with identifying a sufficiently liberal response, do not mean that centralized environmental regulation is an effective means of addressing environmental concerns. Just as the creative application of property rights principles to fisheries has facilitated the sustainable management of marine fisheries, it is possible that the creative application of similar principles to pollution problems could yield more effective and equitable solutions, but this case has yet to be made.

Climate Liberalism

As noted, significant progress has been made in addressing many environmental problems, often as a result of embracing or allowing liberal institutions to operate. Understanding how the liberal institutions of property rights and markets have facilitated environmental progress is essential to replicating these successes and avoiding future environmental failures. Yet, as also noted, the liberal intellectual tradition has been less helpful at informing (let alone solving) other environmental problems, particularly those related to large-scale pollution problems and climate change. This volume aims to help fill this gap.

Individual rights are often the starting point for classical-liberal analysis, so the volume appropriately enough begins with a consideration of how to address pollution within a framework of natural rights. In "Pollution and Natural Rights," political theorist Billy Christmas considers the lessons of natural rights theory for pollution control and suggests how natural rights can be recognized and protected without adopting rules that are overly stringent.[53] Daniel Cole, on the other hand, is less sanguine than Christmas about the potential for liberal principles to yield a workable approach to large-scale pollution problems, such as climate change, particularly if grounded in libertarian conceptions of individual rights.[54]

The collective nature of climate change is part of what makes it such a wicked problem to address, particularly within a rights framework. Nonetheless, courts have considered (and are considering) rights-based litigation seeking to prevent or compensate for the harms of climate change. David Dana surveys some of this litigation and how American courts in particular have considered such claims.[55] Catherine Sharkey complements this analysis by considering the information-forcing role of rights-based litigation, including common-law public nuisance, and how such litigation has evolved, and may still evolve, to fill regulatory voids in a decentralized fashion.[56]

While courts may be able to address some pollution problems through the resolution of rights-based claims, Dan Shahar argues that such efforts will be insufficient to adequately address larger scale problems, such as climate change, and will require the addition of legislative and regulatory measures.[57] Such interventions need not be illiberal, however, as such division of responsibility is consistent with liberal conceptions of the separation of powers.

One particular challenge for liberal, rights-based approaches to large-scale pollution problems is the mismatch between existing property rights and the scale of landscape or atmospheric environmental problems. This mismatch problem is pervasive, suggest Monika Ehrman and Karen Bradshaw, and calls for a multidimensional model of property that can account for a necessarily overlapping constellation of rights across the landscape and ecological resources.[58]

One way to address broad-scale problems is to aggregate interests so that they may be resolved more efficiently or effectively through the judiciary or other dispute resolution systems. Class-action litigation is one such aggregation tool that Brian Fitzpatrick suggests classical liberals are too quick to overlook.[59] Compared with the alternative of centralized administrative regulation, Fitzpatrick argues that class-action litigation is more consistent with traditional liberal values. A second aggregation tool, though one less immediately applicable to climate change, is the use of firms to account for environmental effects.[60]

Centralized administrative regulation often threatens to constrain the dynamic and evolutionary nature of free economies and liberal societies. This is particularly true of "precautionary" regulatory measures that seek to constrain economic activities when there is a risk of significant economic impact. John Thrasher argues that because dynamism is so essential to liberal society, precautionary approaches should be rejected in favor of a "default permission rule," even when environmental consequences are on the table.[61]

One problem with precautionary regulation is that it may chill innovation and foreclose the development of systems and measures to address environmental threats, including those posed by climate change. Yet dynamic markets may often lead to the development of risk-management techniques and welfare-enhancing measures that make environmental threats more manageable. Andrew Morriss explores one such possibility, focusing on insurance and alternative risk transfer methods that can be used to help disclose and account for environmental risks, including those posed by climate change.[62]

While the conventional approach to environmental risks is the adoption of administrative regulations, liberal economists have long been interested in alternative interventions, including the use of fiscal measures. In that tradition, Ed Dolan makes a classical-liberal case for a carbon tax, drawing on Locke, Hayek, and other liberal thinkers.[63] While Dolan would argue that carbon pricing is superior to alternative approaches, Mark Pennington is skeptical that there is any empirical basis upon which such a claim can be made.[64] From a robust political economy perspective, Pennington suggests, it is impossible to evaluate the relative effectiveness of different policy approaches given the global nature of climate change. Despite the problems that Pennington identifies, Mark Budolfson argues that there is nonetheless a liberal argument to be made for pricing carbon consistent with a "social cost of carbon" model, and that epistemic humility should not be a bar to such policy interventions.[65]

Conclusion

Although relatively few classical-liberal thinkers have focused on environmental concerns, the application of classical-liberal principles has helped to address some significant environmental concerns. Competitive markets have encouraged more sustainable resource use and the creative extension of property rights have fostered stewardship and conservation. If classical-liberal principles are to help guide environmental policy in the future, greater attention must be paid to how such principles can apply to more difficult environmental concerns and inform the evaluation of policy proposals.

Large-scale pollution problems, such as climate change, defy easy solutions and challenge the classical-liberal emphasis on limited government and decentralized authority. But just as it was possible to create property rights in marine fisheries, it may be possible to extend liberal institutions into new domains and develop solutions to such environmental challenges consistent with classically liberal principles. It is my hope that this volume contributes to that endeavor.

Notes

1. For a compendium of human progress along these (and other) lines, see Marian L. Tupy and Gale L. Polley, *Superabundance: The Story of Population Growth, Innovation, and Human Flourishing on an Infinitely Bountiful Planet* (Washington, DC: Cato Institute, 2022).
2. For an overview of climate change science, see Intergovernmental Panel on Climate Change, *Climate Change 2021: The Physical Science Basis, Contribution of Working Group I to the Sixth Assessment Report of the Intergovernmental Panel on Climate Change* (Cambridge: Cambridge University Press, 2021). For an assessment of climate science by a prominent libertarian science writer, see Ronald Bailey, "Climate Change: How Lucky Do You Feel?" *Reason*, January 2020, https://reason.com/2019/12/01/climate-change-how-lucky-do-you-feel/.
3. Richard J. Lazarus, "Super Wicked Problems and Climate Change: Restraining the Present to Liberate the Future," *Cornell Law Review* 94, no. 5 (2009): 1159. Lazarus suggests climate change is actually a "super-wicked" problem because of three factors: (1) it becomes more difficult to address over time, (2) those with the greatest ability to address climate change have the least incentive to do so, and (3) there is no institution capable of addressing climate change on the global scale at which it occurs. Ibid., 1159–1161.
4. Much classical liberal work on environmental problems embraces the moniker "free market environmentalism." Examples of scholarship in this vein include the following: Terry L. Anderson and

Donald R. Leal, *Free Market Environmentalism for the Next Generation* (New York: Palgrave Macmillan, 2015); Terry L. Anderson and Gary D. Libecap, *Environmental Markets: A Property Rights Approach* (Cambridge: Cambridge University Press, 2014); Terry Anderson and Donald R. Leal, *Free Market Environmentalism*, rev. ed. (New York: Palgrave, 2001); Fred L. Smith, Jr., "Markets and the Environment: A Critical Reappraisal," *Contemporary Economic Policy* 13, no. 1 (1995); Fred L. Smith, Jr., "A Free-Market Environmental Program," *Cato Journal* 11, no. 3 (1992).

For some of this author's efforts in this vein, see Jonathan H. Adler, "Conservative Principles for Environmental Reform," *Duke Environmental Law and Policy Forum* 23, no. 2 (2013); Jonathan H. Adler, "Free and Green: A New Approach to Environmental Protection," *Harvard Journal of Law & Public Policy* 24, no. 2 (2001); Jonathan H. Adler, ed., *Ecology, Liberty & Property: A Free Market Environmental Reader* (Washington, DC: Competitive Enterprise Institute, 2000).

5. There are some notable exceptions. These include Matt Zwolinski, "Libertarianism and Pollution," in *The Routledge Companion to Environmental Ethics*, eds. Benjamin Hale and Andrew Light (New York: Routledge: 2015); Dan Shahar, "Justice and Climate Change: Toward a Libertarian Analysis," *The Independent Review* 14, no. 2 (2009); Edwin G. Dolan, "Science, Public Policy, and Global Warming: Rethinking the Market-Liberal Position," *Cato Journal* 26, no. 3 (2006).
6. J.B. Ruhl, "Regulation by Adaptive Management—Is It Possible?" *Minnesota Journal of Law, Science & Technology* 7 (2005): 21.
7. For examples, see Jonathan H. Adler, "Dynamic Environmentalism and Adaptive Management: Legal Obstacles and Opportunities," *Journal of Law, Economics & Policy* 11, no. 2 (2015): 142–143. See also John Copeland Nagle, "Green Harms of Green Projects," *Notre Dame Journal of Law, Ethics & Public Policy* 27, no. 1 (2013): 59–104; Frank B. Cross, "Paradoxical Perils of he Precautionary Principle," *Washington & Lee Law Review* 53, no. 3 (1996): 851–928.
8. Richard B. Stewart, "United States Environmental Regulation: A Failing Paradigm," *Journal of Law & Commerce* 15, no. 2 (1996): 587.
9. As Louis Michael Seidman observes, "Unfortunately for libertarians, domestication is impossible because the externalities problem arises in every case. Every transaction produces externalities, if only because some outsiders do not want the transaction to occur." Louis Michael Seidman, "Seven Problems for Classical Liberals," in *The Cambridge Handbook*

of Classical Liberal Thought, ed. M. Todd Henderson (Cambridge: Cambridge University Press, 2018), 270–271. It is precisely for this reason that Ronald Coase rejected the notion that the existence of an externality was sufficient to justify government intervention. Rather, he argued "The ubiquitous nature of 'externalities' suggests to me that there is a prima facie case against intervention." R. H. Coase, *The Firm, the Market and the Law* (Chicago: University of Chicago Press, 1998), 26.

10. As economist Paul Heyne observed, "Ardent environmentalists need to discover and acknowledge that the same limitations which made economic central planning impossible will make it impossible to establish a comprehensive system of central environmental planning." Paul Heyne, "Economics, Ethics, and Ecology," in *Taking the Environment Seriously*, eds. Roger Meiners and Bruce Yandle (Lanham: MD: Rowman & Littlefield, 1993), 47.
11. As Fred Smith has argued, "Rather than viewing the world in terms of market failure, we should view the problem of externalities as a failure to permit markets and create markets where they do not yet—or no longer—exist." Fred L. Smith, Jr., "Conclusion: Environmental Policy at the Crossroads," in *Environmental Politics: Public Costs, Private Rewards*, eds. Michael Greve and Fred L. Smith, Jr. (New York: Praeger, 1992): 192.
12. Matthew E. Kahn and John G. Matsusaka, "Demand for Environmental Goods: Evidence from Voting Patterns on California Initiatives," *Journal of Law & Economics* 40, no. 1 (1997): 137.
13. This is one of the central insights of F.A. Hayek. See F. A. Hayek, "The Use of Knowledge in Society," *American Economic Review* 35, no. 4 (1945): 519–530.
14. *See* Richard L. Stroup, *Eco-nomics: What Everyone Should Know about Economics and the Environment* (Washington, DC: Cato Institute, 2003), 74-75.
15. Seth W. Norton, "Property Rights, the Environment, and Economic Well-Being," in *Who Owns the Environment?* eds. Peter J. Hill and Roger E. Meiners (Lanham, MD: Rowman & Littlefield, 1998), 37, 51.
16. Carrie B. Kerekes, "Property Rights and Environmental Quality: A Cross-Country Survey," *Cato Journal* 31, no. 2 (2011), 315.
17. Christopher Andrew Hartwell and Don L. Coursey, "Revisiting the Environmental Rewards of Economic Freedom," *Economics and Business Letters* 4, no. 1 (2015): 41.
18. Tupy and Polley, *Superabundance*.

19. The ecological record of socialist and communist countries is particularly bleak. Peter J. Hill, "Environmental Problems Under Socialism," *Cato Journal* 12, no. 2 (1992): 321–335; Mikhail Bernstam, *The Wealth of Nations and the Environment* (London: Institute for Economic Affairs, 1991).
20. Andrew McAfee, *More from Less: The Surprising Story of How We Learned to Prosper Using Fewer Resources—and What Happens Next* (New York: Scribner, 2019).
21. Garrett Hardin, "The Tragedy of the Commons," *Science* 162 (December 13, 1968): 1243. While Hardin is rightly credited with popularizing the idea of the "tragedy of the commons," he was far from the first to observe or analyze this dynamic. Hardin was likely influenced by fishery economists who diagnosed the commons problem in marines fisheries in the 1950s. See, for example, H. Scott Gordon, "The Economic Theory of a Common Property Resource: The Fishery," *Journal of Political Economy* 62, no. 2 (1954); Anthony Scott, "The Fishery: The Objectives of Sole Ownership," *Journal of Political Economy* 63, no. 2 (1955). Yet the basic insight goes back at least as far as Aristotle, *Politics* §1261b ("[T]hat which is common to the greatest number has the least care bestowed upon it.").
22. Robert J. Smith, "Resolving the Tragedy of the Commons by Creating Private Property Rights in Wildlife," *Cato Journal* 1, no. 2 (1981): 456 ("Wherever we have exclusive private ownership, whether it is organized around a profit-seeking or nonprofit undertaking, there are incentives for the private owners to preserve the resource."). *See also* David Schmidtz, "The Institution of Property," in *The Common Law and the Environment*, eds. Roger E. Meiners and Andrew P. Morriss (Lanham, MD: Rowman & Littlefield, 2000).
23. Harold Demsetz, "Toward a Theory of Property Rights," *American Economic Review* 57, no. 2 (1967). It is important to note that Demsetz' claim is not that *every* private landowner will act in this fashion, just that the incentives of ownership are such that the typical landowner will act in this fashion. As is true in all contexts, the behavior of specific individuals will vary, with some taking greater or lesser actions to maximize the present value of the property in question. Those property owners which do the best job of estimating likely future income streams are then rewarded in the marketplace with greater property values.

24. As Anthony Scott observes, "No one will take the trouble to husband and maintain a resource unless he has a reasonable certainty of receiving some portion of the product of his management; that is, unless he has some property right in the yield." Scott, "The Fishery," 63.
25. *See*, e.g., Fred L. Smith Jr., "Reappraising Humanity's Challenges, Humanity's Opportunities," in *The True State of the Planet*, ed. Ronald Bailey (New York: Free Press, 1995), 379.
26. Jonathan H. Adler, "Poplar Front: The Rebirth of America's Forests," in *Ecology, Liberty & Property: A Free-Market Environmental Reader*, ed. Jonathan H. Adler (Washington, DC: Competitive Enterprise Institute, 2000); Roger J. Sedjo, "Forests: Conflicting Signals," in *The True State of the Planet*, ed. Ronald Bailey (New York: Free Press, 1995); Roger A. Sedjo and Douglas MacCleery, "Sustainable Forests in America?" in *Perspectives on Sustainable Resources in America*, ed. Roger A. Sedjo (Washington, DC: Resources for the Future, 2008).
27. Richard J. Agnello and Lawrence P. Donnelly, "Property Rights and Efficiency in the Oyster Industry," *Journal of Law & Economics* 18, no. 2 (1975); Richard J. Agnello and Lawrence P. Donnelly, "Prices and Property Rights in the Fisheries," *Southern Economic Journal* 42, no. 2 (1975).
28. Ike Sugg and Urs Kreuter, *Elephants and Ivory: Lessons from the Trade Ban* (London: Institute for Economic Affairs, 1994); Randy T. Simmons and Urs P. Kreuter, "Herd Mentality: Banning Ivory Sales Is No Way to Save the Elephant," *Policy Review*, Fall 1989, 46.
29. As Andrew Morriss notes, "Markets are far from perfect, of course. But, critiques of markets in general, and critiques of water markets in particular, often conflate dissatisfactions with human nature or other features of society with problems in the market." Andrew P. Morriss, "Real People, Real Resources, and Real Choices: The Case for Market Valuation of Water," *Texas Tech Law Review* 38 (2006): 974–975.
30. *See* Martin D. Smith, "The New Fisheries Economics: Incentives Across Many Margins," *Annual Review of Resource Economics* 4 (2012): 379, 380 ("The story of fisheries economics could be distilled into diagnosing the commons problem and offering a solution to it.").
31. Christopher Costello, Steven D. Gaines, and John Lynham, "Can Catch Shares Prevent Fisheries Collapse?" *Science* 321 (2008): 1678–1681.

32. Christopher Costello, John Lynham, Sarah E. Lester, and Steven D. Gaines, "Economic Incentives and Global Fisheries Sustainability," *Annual Review of Resource Economics* 2 (2010): 299–318.
33. Trevor A. Branch, "How Do Individual Transferable Quotas Affect Marine Ecosystems?" *Fish and Fisheries* 10 (2008): 39; see also Jonathan H. Adler and Nathaniel Stewart, "Learning How to Fish: Catch Shares and the Future of Fishery Conservation," *UCLA Environmental Law & Policy Review* 31, no. 1 (2013).
34. Jonathan H. Adler, "Water Rights, Markets, and Changing Ecological Conditions," *Environmental Law* 42, no. 1 (2012): 93–113; Morriss, "Real People, Real Resources."
35. Jedidiah Brewer, Robert Glennon, Alan Ker, and Gary Libecap, "Transferring Water in the American West: 1987-2005," *University of Michigan Journal of Law Reform* 40, no. 4 (2007): 1042.
36. Terry L. Anderson and Pamela Snyder, *Water Markets: Priming the Invisible Pump* (Washington, DC: Cato Institute, 1997), 120.
37. Terry L. Anderson and Donald R. Leal, *Enviro-Capitalists: Doing Good While Doing Well* (Lanham, MD: Rowman & Littlefield, 1997), 94–95.
38. Adler, "Water Rights"; Jonathan H. Adler, "Water Marketing as an Adaptive Response to the Threat of Climate Change," *Hamline Law Review*, 31 no. 3 (2008): 729–754.
39. Hayek, "The Use of Knowledge."
40. K. Duncan et al., "North America," in *Climate Change 2001: Impacts, Adaptation, and Vulnerability: Contribution of Working Group II to the Third Assessment Report of the Intergovernmental Panel on Climate Change* (Cambridge: Cambridge University Press, 2001): 735, 748.
41. Indeed, increases in the security of property rights in land can correlate with increases in air pollution in the absence of corresponding efforts to protect property rights from such violations. Kerekes, "Property Rights and Environmental Quality," 315.
42. The precise nature of the property rights violation would be dependent upon the contours of the property rights at issue.
43. Anderson and Leal, *Free Market Environmentalism*, 132.
44. Coase, "Problem of Social Cost."
45. These concerns are laid out in further detail in Jonathan H. Adler, "Is the Common Law the Free Market Solution to Pollution?" *Critical Review* 24, no. 1 (2012).

46. For public choice analyses of environmental policies, and examples of how environmental regulations may be manipulated or compromised by special interest pleading, see Terry L. Anderson ed., *Political Environmentalism: Going Behind the Green Curtain* (Stanford, CA: Hoover Institution Press, 2000); Jonathan H. Adler, "Rent-Seeking Behind the Green Curtain," *Regulation* 19, no. 4 (1996); Michael S. Greve and Fred L. Smith, Jr., *Environmental Politics: Public Costs, Private Rewards* (New York: Praeger, 1992).
47. Adler, "Is the Common Law."
48. For example, climatologists Patrick J. Michaels and Robert C. Balling, Jr., two prominent warming "skeptics," predicted that there will be 0.65–0.75 degrees Celsius warming by 2050. Patrick J. Michaels and Robert C. Balling, Jr., *The Satanic Gases: Clearing the Air about Global Warming* (Washington, DC: Cato Institute, 2000), 210–211. They also predicted a warming-induced sea-level rise of 5–11 inches over the next century. Ibid., 162.
49. Jonathan H. Adler, "Taking Property Rights Seriously: The Case of Climate Change," *Social Philosophy & Policy* 26, no. 2 (2009).
50. Robert Mendelsohn, Ariel Dinar, and Larry Williams, "The Distributional Impact of Climate Change on Rich and Poor Countries," *Environment and Development Economics* 11, no. 2 (2006).
51. Adler, "Taking Property Rights Seriously."
52. Bob Litterman, "What Is the Right Price for Carbon Emissions?" *Regulation* 36, no. 2 (2013): 38–43.
53. Billy Christmas, "Pollution and Natural Rights," this volume.
54. Daniel H. Cole, "Do Libertarians Have Anything Useful to Contribute to Climate Change Policy?" this volume.
55. David Dana, "Climate Change Adaptation through the Prism of Individual Rights," this volume.
56. Catherine M. Sharkey, Common Law Tort as a Transitional Regulatory Regime; A New Perspective on Climate Change Litigation," this volume.
57. Dan C. Shahar, "Libertarianism, Pollution, and the Limits of Court Adjudication," this volume.
58. Monika Ehrman and Karen Bradshaw, "Complexities of Climate Governance in Multidimensional Property Regimes," this volume.
59. Brian T. Fitzpatrick, "Climate Change and Class Actions," this volume.
60. Jonathan H. Adler, "Nature and the Firm," this volume.
61. John Thrasher, "Permission, Prohibition, and Dynamism," this volume.

62. Andrew P. Morriss, "Market Solutions to Large Number Environmental Problem-Induced Changes in Risk Distributions," this volume.
63. Ed Dolan, "A Classical Liberal Case for Target-Consistent Carbon Pricing," this volume.
64. Mark Pennington, "Climate Change, Political Economy, and the Problem of Comparative Analysis," this volume.
65. Mark Budolfson, "The Social Cost of Carbon, Humility and Overlapping Consensus on Climate Policy," this volume.

Pollution and Natural Rights

Billy Christmas

Introduction

On the heels of a report in 2017 the UN announced that the British government's failure to mitigate pollution was a *human rights violation.* The rather more difficult task of formulating precisely what the violated rights were, how they should be remedied, by whom, and most importantly, *why*, are left out. Nonetheless, the general claim that high levels of pollution can be regarded as a human rights violation should be welcomed, even in such a vague and haphazard way. Often, pollution is regarded as an environmental issue—a problem that faces the environment, and that if we care about the environment, we

I am grateful to the other contributors to this volume, especially Jonathan Adler, for their very helpful comments and suggestions on an earlier draft of this essay.

B. Christmas (✉)
King's College London, London, UK
e-mail: billy.christmas@kcl.ac.uk

J. H. Adler (ed.), *Climate Liberalism*, Palgrave Studies in Classical Liberalism,
https://doi.org/10.1007/978-3-031-21108-9_2

should care about too. Pollution is often regulated under "environmental protection." However, from the perspective of rights, pollution is an interpersonal problem—when we impact one another's space in harmful ways, we harm one another. Human beings, as such, have rights that their bodies and physical environments not be interfered in certain ways. Very often, then, failure to protect the environment from people is a failure to protect people from one another.

The idea of "human rights" as deployed by the United Nations stems from what the early modern Protestant natural law theorists of Western Europe called *natural rights*.[1] These rights are natural insofar as we have them in virtue of our nature; in other words: as humans.[2] The alternative would be the view that we have rights only because we live under institutions that recognise them, or we only have rights because we believe that we have them and act like we have them. Natural rights are real and apply always and everywhere to all persons. They are not merely conventions or human contrivances.

This essay outlines the way in which natural rights theory regards pollution and on what basis it will say that pollution violates persons' rights. I will also outline some key objections that have been made to its response to pollution, but argue that these are surmountable. More serious ontological objections can be made—unfortunately I present these without offering any way out for the natural rights theorist. However, these ontological objections also raise surprisingly interesting insights about the relationship between how we determine that a purported pollutant constitutes a rights violation, and the social context in which it is purported to occur. The self-understanding that social groups have of their activities is essential to determining what counts as a harmful interference in them, and therefore what counts as rights-violating pollution. These are important issues worthy of discussion even in lieu of a full solution.

Natural Rights Theory

Natural rights theory is a framework for determining the normative modalities of human actions and of socioeconomic and political institutions. The rights of natural rights theory are *natural* inasmuch as they are purported to be held by persons in virtue of their humanity as such, and not in virtue of them being recognised by social norms or legal institutions, nor by their being believed to exist by other persons. The existence of natural rights are moral facts about human beings,[3] not about human beliefs or institutions. Natural rights correlate with natural duties. Failure to fulfil these duties constitute a violation of the correlative rights; any action which constitutes either an omission with is required by a natural duty, or an action which is prohibited by one, is wrongful.[4] Actions can be assorted into three categories depending on where they fall within a set of rights: compulsory, prohibited, or permissible. It is compulsory that one *not* violate the rights of others, which means it is *prohibited* to engage in any action that does so. All actions that are within one's own rights are permissible and not compulsory,[5] except where those actions constitute the fulfilment of duties to others (in which case they are both permissible and compulsory.[6] Through this triune enumeration of human actions, theories of natural rights also tell us which socioeconomic institutions are permitted, prohibited, and compulsory, in virtue of speaking to the normative status of the actions that those institutions in turn render permissible, prohibited, and compulsory. Socioeconomic institutions that make it compulsory to engage in actions that were rights violations are ruled out by the relevant theory of rights. In guiding human action, theories of natural rights thereby also tell us what kinds of systems of property, contracts, and torts are justified based on the kinds of human interactions they permit, prohibit, and require.

A theory of natural rights is not an exhaustive morality. It is not tuned to make pronouncements on the relative goodness of all actions. It is, however, exhaustive about actions that involve the use of force and coercion. Since natural rights are, as such, *legitimately enforceable*, their contents are ultimately authorisations to use force or coercion in order to compel compliance, defend oneself or others from a rights violation, or perhaps to punish a rights violation and/or extract compensation after the fact. When you have a right to take a particular action (it being permissible), this does not necessarily mean it is good or right for you to take that action, but it does mean that it would be wrong for another to coercively interfere with you taking that action. The triune enumeration of actions tells us which actions people are morally free to take, but it does not tell us what we ought to do *within* that protected sphere of freedom. There is of course a lot more to morality than when it is morally acceptable to use force or coercion. There is a lot more to being a good or righteous person than simply doing things that you have a right to do.

Given that natural rights lay out the legitimate use of force in society, natural rights theorists often regard the legitimate activities of political institutions such as the state as being determined by the contours of natural rights. Broadly speaking there are two families of accounts of the relationship between natural rights and the political institutions that interpret and enforce them. One is associated with John Locke and says that political institutions are strictly the product of a voluntary agreement among members of a society to defer the enforcement of their natural rights to a common third party—the state.[7] The state's authority to use force in the administration of rights, then, is strictly with the consent of its citizens. The other account is associated with Immanuel Kant and says that political institutions that interpret and enforce rights are automatically authorised on the grounds that without them our rights would be in peril, and the only way we can really respect one another's rights is in and through some sort of civil authority like a state.[8]

What is common between the two views, however, is the constraints that natural rights place on our conduct—the spheres of freedom they assign to each person—determine how the state must be structured. On both accounts, theories of natural rights draw on the ethics of interpersonal interaction at the micro level to inform a normative theory of political institutions at the macro level.

So much for the place of natural rights within moral philosophy and political theory. Substantively speaking, then, what exactly do we have natural rights *to*?

The central distinction between the kinds of natural rights we have are *original* rights and *acquired* rights. Original rights are the ones no one else could have but us—we are born with them and have them in virtue of who/what we are, and not in virtue of things we (or others) have done to acquire them. These are our rights over our bodies and our freedom of action. Acquired rights are rights that could in principle belong to anyone; those who ultimately have them do so because they do or have done something in particular to acquire them. These are property rights in particular objects, or rights that others perform on particular contractual obligations.

There are broadly two ways of acquiring rights: original acquisition (acquiring them ex nihilo) or in transfer (acquiring them from someone else who had them before).[9] The acquisition of rights is regarded as an entailment of an exercise of one's original rights. In the Lockean tradition, one's original rights are those of self-ownership—one has an exclusive right over one's own body and mind—and one uses these rights to acquire others. In making use of hitherto unused external resources in the world, one puts something of oneself into them and joins them to oneself in a way that makes them one's own also. Property rights then are really functions of the rights we have over ourselves extended into extra-personal objects.[10] In the Kantian tradition, the distinction is between innate rights and acquired rights. "Innate" rights are the rights one has to control one's own person and acquired rights are those rights we have to extra-personal objects. Although Kantians do not believe you can acquire fully-fledged property rights in the external world through a mere exercise of one's innate bodily rights as Lockeans do, they nonetheless regard them as being essential to the full realisation of personal freedom.[11] Whilst Lockeans believe individuals can unilaterally acquire property rights in resources through use (and subsequently transfer them to others), Kantians believe that fully-fledged property rights require a "civil condition" or, in other words, a legal system laid down by the state that is authorised to create rights and correlative duties with regard to external resources.

Natural rights theorists also believe that we have moral powers over our original or innate rights as well as our acquired rights. These powers enable us to transfer rights to others when we choose to do so. We can therefore exchange property as well as create obligations to others on a consensual basis. Our natural rights therefore give normative force to familiar areas of private law such as torts, property, and contracts.

In both views, our rights over ourselves, our claim to have rights over external objects, as well as the power to create consensual contractual relationships, is an expression of our moral equality. None of us come into the world naturally superior or inferior to one another. We each have rights over ourselves that none other can have over us, and none of us come into the world with natural claims over external resources—in contra-distinction to those who held the Divine Right of Kings gives monarchs natural claims over territories, and thereby natural authority over their subjects who inhabit those territories.[12]

Pollution and Natural Rights

Legislation and regulation that constrains activities that generate pollution reduces the things that would-be polluters are at liberty to do with their bodies and property. At first blush, one might think that natural rights must protect polluters absolutely in doing as they wish with their private property, and that inasmuch as limiting pollution limits the freedom of polluters, this is something they have a right against. This is not necessarily so; and indeed, often not so.

The activities of axe-murderers also involve the killers' use of their own property as they wish, but we would not for a moment think that prohibiting axe-murder would be an infringement of the murderer's personal liberty or of his private property rights in his axe. The liberties someone has to make use of that which they own do not include actions that would violate the rights others have over their person and property.[13] The rights others have to their property correlate with duties we have over how we conduct our activities; only that action-space in which we can enter without violating a duty to others is where our liberties are enumerated.

Many forms of pollution damage or constrain existing uses that others make of their own property and persons; or if they do not actually impose such damage or constraint, they impose significant risk thereof.[14] They do so without the consent of those on the receiving end. If *A* runs toxic waste materials out of her factory into a river, and that river water, upon contact with *B*'s land, destroys the mineral composition of the soil, then *A*'s actions have invaded the property of *B*, just as if *A* had trespassed onto *B*'s land and damaged it by her own hand. Forms of pollution such as these are relatively straightforward torts since they non-consensually damage the victim's property, interfering with her activities, and imposing costs upon her. *B* therefore has a right that *A*'s polluting activities immediately cease (and if necessary, the use of force or coercively backed threats of the law are legitimate), and a right to restitution—that *A* make her whole again whether in cash or in kind for the damages to her livelihood that she has experienced. The polluter must pay, but the polluter must also stop polluting. This is what it means to regard pollution as a rights violation.[15]

Cessation and restitution is not the only just solution, however. *B* might be happy to put up with a degree of toxic waste on her property as long as *A* pay her enough to cover the cost this imposes on her. Given that the rights over the river are *B*'s to do with as she wishes, she is free to negotiate terms with *A* over how the river is to be used. If *A* is willing to pay *B* a great deal in exchange for a small amount of pollution, and *B* is happy with this arrangement, then they can enter a positive-sum agreement rather than one of mere reparation.

There is an important difference though between when restitution is made for a rights violation done, and when a fee is paid in exchange for a privilege. When restitution is paid, it does not make the previous rights-violation right—it does not undo the wrongdoing. It merely (imperfectly) makes the victim whole again; it does not reverse her victimisation. However, when a fee is paid in exchange for some agreed level of pollution on a contractual basis, no wrongdoing occurs because the receiver of the pollution has agreed in advance and is receiving payment at a rate they deem worth the cost.[16]

Such cases are simple enough. Other forms of pollution prove more complex. In particular, when the number of individuals causing the pollution increases. An example would be when a child is hospitalised with asthma due to pollution from a significant increase in traffic in her neighbourhood. Each individual driver has made a negligible contribution to the violation of the child's bodily rights. Indeed, if all but one driver were stripped from the equation, the rights violation would not have even occurred. Nonetheless, the outcome of the collective, albeit unintended, action of local motorists was that the child's rights were violated, and they violated them. In this instance, each individual motorist is responsible for a small part of the significant rights violation and therefore the liability to pay restitution is proportional to that. Moreover, motorists must be stopped from driving in the area to the degree that it presents such serious health problems or risk of such problems (coercively if necessary). Hence laws minimising excessive car travel in residential areas are necessary to protect certain rights of children. Ascertaining who owes how much compensation in such cases are incredibly difficult, and may not always be worth the cost. Indeed, this problem also afflicts any possibility of polluters and victims entering an agreement over acceptable pollution levels and ongoing compensation to be paid. However, the fact that a rights violation is occurring is sufficient to justify the coercive imposition and enforcement of legislation limits the activity that has this effect, in principle.

A more problematic complexity that can arise is when the chain of causal liability becomes harder to identify between rights violator and rights violation. This can occur simply because of the complex nature of the natural causality in question. For example, we know what kinds of activities cause acid rain, and we know who engages in these activities, but it is impossible (at present levels of technology) to know who in particular is responsible for acid rain in a particular place.[17] Global climate change caused by global CO_2 emissions generate similar problems on an even more serious scale. Such complexities present unsurmountable problems for identifying parties who must pay restitution or who must enter negotiations to price the externality. But again, such a problem does not raise any justificatory problem for coercively prohibiting the polluting activity.

Recall that the liberties one has to make use of one's own property depend upon the rights that others have to their property and one's correlative duties not to interfere with them. In the above examples, the victim is located within their rights at time t_1 and the pollution enters that location at t_2 and therefore the respective polluters are the ones invading the rights of the respective victims. However, we can imagine cases where something of the reverse takes place. Imagine A purchases an apartment property in an area that is mostly used for car parks, warehouses, and late night clubbing. The clubs set up in this area because it was the only part of the city where they could be loud without invading anyone else's established right to have peace and quiet at night in their homes. *B*'s nightclub makes a significant amount of noise on Friday and Saturday nights that stops *A* from being able to sleep or really carry on anything like any normal activities in her new apartment. *A* believes that her rights are being violated—the noise pollution interfering with her use of her home and indeed her body.

She is mistaken. When *B* acquired her property and began using it as a nightclub, she acquired an easement to emit noise pollution in the surrounding area. She had the liberty to do this at that time because she had no duty not to. No one in the surrounding area had acquired a right to peace and quiet. Like the original acquisition of private ownership rights, easement rights can be acquired simply by initiating an activity that does not interfere with the rights of anyone else. And indeed, it did not, so *B* acquired the noise pollution easement. *A* believed (we assume) she was purchasing the mirror image of that pollution easement—a right that the property be subject to peace and quiet every night—but that right was not part of the bundle she acquired in the purchase. If she wants this right, she would have to purchase it from *B*.[18] Of course, a residential developer in the area could have bundled this right with the apartment, but the developer would have to have acquired it from whoever currently held it, in this case, *B*.

Therefore, it is possible that individuals can have a right to engage in what we would otherwise refer to as pollution if they initiated these activities at a time when the activities did not invade anyone else's rights.[19] Newcomers have to work around them and not the other way around. In principle, one could acquire the right to deposit certain materials in a river at t_1 and anyone who decides at t_2 they actually want the river to have a different disposition—one that would benefit them and their property more—they have no right to this. However, it is important to note that the impact of any given activity can change or compound over time even if the activity itself remains unchanged. So, an activity that is non-rights violating at t_1 might become rights violating at t_2 (say, the material deposited is not naturally decomposed of filtered out of the river, but builds up). But it does not become a rights violation simply in virtue of a third party arriving and wishing to make some new use of the river that is incompatible with its current use.

The Efficiency Objection

The first challenge to this approach to consider is that it unduly hinders economic efficiency. Ronald Coase observes that the problem of negative externalities like pollution is not that they occur per se, but that they occur at a level beyond the level at which society is compensated for them by the positive externalities of the polluting activity.[20] For any given polluting activity, it may be the case that stopping it altogether so that the cost imposed by the pollution is reduced to zero will also reduce the social benefits produced by the activity itself to zero. What is important, then, is not to stop all activity that is prima facie harmful to others, but rather to establish at what margin the harm is worth the benefit.

A natural rights approach to pollution mitigation makes it *possible* for negative externalities to be priced but *only if* the party falling prey to pollution chooses to enter a bargain rather than availing themselves of their right to force cessation of the polluting activity. The problem—according to students of law-and-economics—is that it does not *require* that the externality but subjected to an interpersonal cost–benefit analysis.

Construing individuals as having rights over physical objects and space in this way means that minor infringements thereof that could aggregately produce substantial social benefits can be blocked by individual right-holders on the basis that its social costs target them in particular. For example, a property owner may object to the noise nuisance caused by a train line running a few miles from their property line. Given that this nuisance occurs, inter alia, on her property without her consent, she has the moral power to stop it from taking place. Those who own the train line can offer some compensation to her in the hope that it induces her to allow it, but she is within her right to decline this even if the compensation—in some interpersonal standard of welfare—more than compensates for the loss of welfare caused by the noise. Indeed, if this rail line were allowed to run trains for travel and freight it would produce enough welfare that the adjacent property owner could indeed be so compensated such that she is made better off than if she stopped it.

Richard Epstein has argued at length that property rights infringements should be legally permissible in cases where it allows activity to take place such that all affected parties are made better off. Either because those subject to the infringement directly benefit from the activity (an underground sewer construction that requires access to her property) or indirectly from that activity producing welfare part of which can be redistributed to her.[21] In welfare terms, infringements that generate Pareto improvements either directly or through Kaldor-Hicks improvements (followed by payment in money or in kind) are said to be thereby permissible.

Epstein and those following this view would argue that this in no way throws rights under the bus because it identifies who is rightfully owed compensation by asking who held the original right. It is just that not all rights are construed as being protected by *enforcement* rules; some are only protected by *compensation* rules. Whilst compensation is always due ex post a rights violation, in this case compensation is taken to be an ex post justification of the original (what looked like a) rights violation.[22] Attenuating the normative boundaries of some rights to liability rules rather than property rules benefits everyone without making anyone worse off and is therein justified.

The natural rights response to this would be the following. The welfare standard required to make such judgements—if available[23]—may not be the appropriate scale of value to determine the contours of rights. Rights are there to enable people to make their own judgements of value and precisely to protect them from being subjected to the standards of another.[24] Therefore, to subject the contours of rights themselves to some comprehensive theory of welfare would reintroduce a problem we were trying to get away from.[25]

Supporters of natural rights may also be inclined to say it is insufficient to prove that a prima facie right is not an all-things-considered right to merely show that it has a social cost. The social cost of keeping rights strong and unattenuated is the cost of recognising the dignity each individual has as a rights holder. Dealing with people on the basis of rights can be incredibly costly. But this is understood by their defenders to be their virtue—that they protect individuals from all subjection, even unto Pareto.[26]

The Liberty Objection

Natural rights are often said to give individuals *strong*, *robust*, *absolute*, or *unconditional* protections against wrongful treatment.[27] From inside the normative perspective of rights themselves, any violation of a right is entirely prohibited, and any action that is not a violation of a right is entirely permissible (inasmuch as one has no enforceable duty not to). The picture of moral space this might paint is one of a binary: total freedom and total unfreedom. You have total freedom within your rights and total unfreedom beyond them. Moral space is a map and your actions are hemmed in by the enclosure of the lines. The problem this raises is that many if not all of the activities we might want such freedom to pursue have repercussions beyond these boundaries even if we would not colloquially regard them as taking place beyond these boundaries. Anything I do with my body effects the surrounding space, anything I do with my home effects the ground beneath, the air above. When we factor in modern means of production and transport, as well as ecological and technological interdependence, we see that this problem effects the standard case not merely the marginal case.

Peter Railton[28] has argued that liberty rights become, in many cases, unexercisable because of the claim-rights everyone else has. For me to merely drive my car within earshot of your house, or to emit a minute amount of CO_2 within your vicinity violates your claim-rights over your own body and property. I did not, presumably, have your consent to impose these physical changes upon your property, and therefore they were unjust rights violations. The rights violation is not to be identified with the *degree* of damage inflicted or economic cost imposed,[29] but simply that they involve the use of physical objects and spaces that another person has the right to control, and that they happened without their owner's consent.[30]

David Sobel argues that the equal and maximal strength of our claim-rights make us reciprocally "hog-tied" and unable to use our liberty-rights. The whole point of having strong claim-rights is to equally protect individual freedom, and yet they seem to hinder it. Such a framework, he argues, is incoherent. Our liberty becomes so strong as to be petrified.[31]

However, natural rights theorists are largely in consensus that this critique only sticks on a very crude articulation of what our rights actually amount to. Eric Mack[32] turns the incoherence charge on its head by arguing that since claim-rights are indeed there to protect the liberty-rights, they should only be made *just* as strong as they need to be to ensure the liberty-rights can be enjoyed, rather than reciprocally rendering them unexercisable. This does not make the claim-rights we have normatively weaker or "non-absolute", it merely specifies them in a way so that they do not grant exclusionary protection where it is not in fact needed to serve liberty.[33] As I have argued elsewhere, essential to this view working out is an account of the specific content of the liberty-rights our property rights include. The liberal approach is for the liberty-rights we have to be ones to the continued engagement of the activities in which we engage, that themselves do not interfere with any pre-existing activities anyone else was engaged in before us.[34] This means our claim-rights to non-interference are just strong enough for us to pursue activities, but not so strong that they exclude us from initiating activities that do not themselves present any hindrance to anyone else.

Conventionality

The problems in the family outlined above, are surmountable. At least, sufficiently so for the applicability of natural rights to be retained in the face of the problem of pollution for those who maintain the truth of its general deontological core. But of course, if one holds an interpersonal metric of optimisable value as paramount, then the disagreement runs deeper and the problem of pollution poses no *special* challenge to natural rights theory from that perspective.

There is however a deeper problem that resides in natural rights theory that I believe comes especially to the fore in the face of the problem of pollution. On one level it is easy to say what our natural rights amount to. To not be defrauded, assaulted, enslaved, or have our property taken, used or damaged without our consent, for example. Or perhaps as a Kantian might say, to have an equal and maximal share of external freedom. Or, as a Lockean might say, our rights of self-ownership and legitimately acquired rights of world-ownership. And so on. But in each particular case for each particular person, we need to have a very concrete account of what in particular these abstractly framed rights amount to. What specific actions by others does it give them the power to enjoin, and what specific actions of their own does it give them the liberty to engage in? The fact that we each have a right to private property does not tell us who owns what (much less, if I own a river, do I have a right that you not alter the soil composition on the adjacent land).

The problem I'm alluding to is not that we may *disagree* over what the concrete content of our rights amount to—this is merely an epistemic problem which is not a problem at all for a *theory* of rights, even if it raises problems at the "non-ideal" level of administration.[35] Rather, the problem is that there may be no concrete *matter of fact* either way. This is an ontological problem—which is very much a problem for a theory of rights!

A common retort to this problem is that we need social conventions to specify the concrete content of our abstract rights.[36] But what exactly does that mean?

There are two ontological levels at which social conventions can operate.[37] The first is as *regulative* rules: these regulate which actions may, or may not, and must be executed within the social group to whom they apply. We can observe which regulative rules a social group has conventionalised by observing which actions take place without sanction, which actions are sanctioned, and how they are sanctioned, etc. Natural rights theory tells us what regulative rules we *should* adopt as a society insofar as it tells us what we as individuals may, or may not, and must do vis-à-vis one another and the external world. The positive adoption of regulative rules that conventionalise our natural rights enshrines them as social or legal rights. But regardless of what regulative conventions societies in fact adopt, the normative demands of natural rights remain. Conventions can be deeply unjust by the lights of our natural rights!

The second level at which social conventions can operate is as *constitutive* rules. These rules do not merely regulate the activity, but they also *define* it. That we transfer property rights in land by signing our names on documents that affirm our consensual transfer of the rights is a constitutive rule. This is what defines the *conveyance of land* in a particular community. The moral move of transferring the right is accomplished by engaging in the physical acts of obtaining these documents with the right words on them and voluntarily signing and filing them. The transfer is constituted by such actions. Now, a theory of natural rights will tell us that we are not permitted to make use of property we have no right to, and that such rights can be transferred voluntarily by the owner to whom they choose. But it does not tell us what concrete actions must take place for such a transfer to have taken place. Rather, it presumes there is some fabric of constitutive rules so that when it stipulates the regulative rules we should adopt, they have some referent. Without some socially intelligible mode of giving consent, the normative demand of voluntary exchange makes no sense.

Even the edict "do not interfere in one another's actions" presupposes and requires some prior constitutive rule for what counts as interference. Do I interfere with your use of your body and home when I light a bonfire in my garden next door? What about if I turn my house into a nightclub? Or a nuclear reactor? We need a thick nexus of constitutive rules in order to categorise one another's activities in ways that enable them to be picked out by regulative rules.[38] This means that the way in which a scheme of natural rights is applied will covary with the social, cultural, ecological, and technological context.

Now we are able to see the deeper problem. In order for our abstract natural rights to be furnished with concrete actionable content, it is necessary that there be a nexus of integrated constitutive rules among the interacting group. In order for regulative rules (like rights) to be applicable—that is, for there to be a fact of the matter about how they apply in any given case, making pronouncements on wrongdoing and remedies, etc.—there must be prevailing constitutive rules to give fixity to the actions they refer to. If the regulative prescriptions of natural rights theories cannot refer, they cannot be normative for us. What we require therefore is a hermeneutically integrated community among whom agreement can be obtained as to how these rights apply, *in principle.* I say in principle because such an agreement does not need to actually be obtained, and will likely be unobtainable due to ignorance and moral failing. Nonetheless, because there is a fact of the matter as to what constitutes this or that action, and such can be demonstrated by observing and pointing to the way in which activity and language proceed in the community, it is at least possible for there to be such agreement where there would not be there hermeneutical integrity.

Now, the conditions that one can expect to yield hermeneutical integrity are precisely those conditions that may be absent between polluters and those harmed by pollution. Hermeneutical integrity means agreement on what things mean. Not agreement in terms of explicit and deliberate assent among persons, but simply agreement in their forms of life.[39] There is an agreement between the way they do things with words and deeds such that identity claims can be made between two different people's actions. Of course, no two actions are qualitatively identical let alone numerically identical. But among the hermeneutically integrated they can recognise what one another is doing because they do things the same way.[40]

To clarify: two people may disagree over what the best use of a pond is, and disagree over whose use of the pond should be prioritised. We cannot both get our way because of the scarcity of the resource in question. It is such inevitable disagreement that gives rise to the deontic need for rights in the first place.[41] However, for them to be subject to rules governing the resource, giving priority to a particular use or user over others in a particular way, it is essential that those users be able to recognise one another's activities in a common way. For them to be able to determinately and credibly say that "that kind of fishing is not permitted", or "you are only allowed to swim, not to wash your clothes," or "you are not permitted to baptise." There must be common standards of meaning assigned to different activities for property rules to apply.[42]

Constitutive rules develop over long periods of time among those who regularly encounter one another through convergence on particular ways of doing things they need to do. Just as common languages emerge, so do common ways of acting. Really, language is just one example of "common meanings"[43] emerging among those who encounter one another and therefore need and eventually converge on a common form of life. Among persons who live, work, compete, play, and worship among one another we can expect such hermeneutical integrity and observe it.

However, pollution occurs when one party's activities unintentionally interfere with another's. Typically, these impacts were unforeseen because the affected party is *far away*. They cannot immediately provide feedback on the consequences of the polluting party's activities, even if they have sufficient knowledge of their providence. So, between polluter and polluted there may be no background of ongoing interaction, cooperation, or indeed common understandings regarding the constitution of one another's activities. Some rights violations offer grounds to presume hermeneutical integrity among the implicated parties. A violation of contractual obligations presupposes some intention to cooperate, some agreement to a contract, etc. An assault on the street presupposes physical proximity. The very nature of pollution itself gives us strong reason to presume a lack of hermeneutical integrity, at least in the most serious, international, and large-scale cases.

I cannot offer a solution to this problem but rather raise it to note, at best, that there are limits to the application of natural rights theory. There are certain social conditions under which deontic categories can apply. However, this does not mean that natural rights turn out to be merely conventional because those social conditions are not that the rights are recognised and enforced, only that their content can be mutually intelligible to those they apply to.

What I frame above as a challenge to natural rights theory can also be constructive in thinking about the problem of pollution. Given that the application of natural rights covary with the social, cultural, ecological, and technological context, and that these factors change over time, the content of natural rights also changes over time.

In a hermeneutically integrated community: changes in social attitudes towards risk, health, and the constitution of activities more generally means it is not a fixed matter which kinds of pollution are and are not actionable. This is an essential insight when structuring regulation that would prohibit would-be actionable pollution ex ante.

Take the example of smoking in public. There was a time when we did not know that second-hand smoke was harmful to our bodies, and if we found it disagreeable psychologically, we nonetheless accepted it as a fact of life for smokers and non-smokers alike. It was not believed to be, nor felt to be an interference in one's daily activities nor a violation of one's bodily integrity. With improving information about what tobacco smoke does to the human body, we now know that it is very harmful. With that knowledge, it began bothering people and being felt as an intrusion into their body, their health, and their daily activities. However, the fact is that second-hand smoke was interfering with people even *before* they knew it. Physical processes were occurring in their bodies that were at odds with the plans that those people had for their bodies (i.e. not getting respiratory diseases, not damaging their lung capacity). Scientific knowledge in this sense is essential to discovering what pollutants are rights-violating *given* the background assumption that seriously damaging people's health interferes with their plans and projects. With the development of scientific knowledge then, we can discover actionable pollutants we were unaware of.

A different example would be the pollution of equine faeces in the streets that all city dwellers experienced as part of their daily life before the advent of the motor car. This was an accepted fact of daily life in those days, exposure to which one tacitly consented to by choosing to live in a city. Of course, if you stepped out into the high street now and the floor was coated in horse faeces, the site and the stench would probably give you quite a shock and you would rightly feel that your (and your neighbours') easement rights to this space had been flagrantly violated. Now that this experience is not constitutive of the activity of *taking a stroll around town*, its imposition onto that activity constitutes an interference. We did not *discover* that animal faeces are repulsive and those poor people in the nineteenth century were having their rights to public space violated on a daily basis. Rather, our social attitudes adapted to the facts of life and transformed what does and does not constitute a rights-violating interference with our use of public space. (Whether or not the pollution, noise, and danger from motor cars hurling past us are better or worse I will leave as a consideration for the reader.)

Indeed, those pollutants that are classed as nuisances will often be so inasmuch as they interfere with "ordinary use" of other's property. What constitutes ordinary use depends upon the prevailing social practices. The constitution of activities is constructed in part by the surrounding environmental constraints including the activities of others. What is ordinary will change with social attitudes, and those changing social attitudes can be driven by changing culture, technology, information. Nowadays in the part of the world I am writing from, people's ordinary use of their homes includes *not* being subjected to the smoke of a brick kiln on their neighbour's land, but there was a time when social practices were not that way configured.[44]

There are three ways in which activities that were not initially pollutants can become pollutants. Through (1) objective discovery (as with tobacco smoking), through (2) social changes (as with equine faces and with kiln smoke). Additionally, through (3) objective changes. There are two forms of (3): When the effects of an activity accumulate in such a way as to have a new affect that was not present before, such as is the case with CO_2 emissions. This is when there is an objective change in the effect of the pollutant due to changing facts about the pollutant itself such as its magnitude. Or when the objective circumstances of the polluted change which alter the character of the externality. For example, carrying an open packet of peanuts around a school in one's bag would have not long ago been assured to be a presumptively innocuous activity, having next to no negative externalities whatsoever. Changes in the human immune system, however, mean that some children have such severe nut allergies they exposure to the aroma of peanuts can carry serious health risks.[45]

The fact that social context matters to how natural rights are to be applied does present limits to the power of the framework for dealing with pollution, as noted above. However, that fact also illuminates important ways in which normativity covaries (or not) with the changing state of social practices and technological development. In the moral, political, and economic analysis of pollution, it is essential to keep at the forefront of one's mind that fixed though morality is, social, ecological, and technological dynamism present enormous variables in how we should respond to pollution. We do not want to generate regulations that presuppose our social attitudes will never change, nor that technology will change our options and give us the luxury of deeming things pollutants we did not in the past. We should also not be too quick to suppose that what constitutes a pollutant in one place (or time) automatically does in another. Local knowledge of constitutive practices is essential to determining wrongdoing in cases of pollution.

The Efficiency Objection, Redux

The first efficiency objection was that natural rights give individuals the right to decline terms of exchange even when such terms appear—from an interpersonal perspective—to promise Pareto improvements. This objection really cuts to the core of the deontic framework which is presupposed by natural rights theory, and whilst the problem of pollution may provoke this objection, it is not a special objection to natural rights theory's approach to pollution. A second efficiency-based objection however cuts not against the internal axiological approach of natural rights, but specifically to the incentives created by their mode of acquisition—and hence does raise a special problem with regard to its application to pollution.

Natural rights theories typically allow that rights in extra-personal resources be acquired through some kind of ostensive *use*. For our rights to protect our liberty from interference, it must be those actions that we in fact initiate in relation to the external world that must be protected. If someone interferes with your use of some hitherto unclaimed resource, then they interfere with your activities, and therefore property rights are acquired through use. Locke himself used the metaphor of *labour-mixing*[46]—when you mix your labour with a resource, it is now intermingled with something only you own (your labour) and therefore the ownership "seeps over into the rest."[47] It is often understood that the metaphor of labour here is standing in to mean that one comes to own the things one puts to *productive use*.[48] Whilst this incentivises socially beneficial economic uses by creating residual claimants to those benefits, it may sometimes do so to an excessive degree that is actually inefficient.

Allowing pollution easements to be acquired through first use incentivises resource exploitation beyond the rate at which resources are actually demanded at the expense of reduced option-values. When actors can acquire pollution easements of various kinds with regard to resources such as water tables, rivers, swathes of land, airspace, etc., simply by being the first to make such resources productive, they will engage in such activities because otherwise, they may miss out on a later opportunity to use the resources in any way at all. Thus, generating a race to the top for making resources productive before others do, simply for the sake of not missing out on them later even if there is not huge value to making use of them now.[49]

The most value that may be derivable from a resource might be some use that is entirely unknown to us now, and we may very well frustrate any future possibility of overusing the resource now. There is a great deal of uncertainty about what value we can derive from present resources in the future, and given that some of these dispositions will be incredibly valuable, efficiency demands some degree of holding resources back for future unknown uses. Only granting property protections to uses of resources that make them ostensibly productive in the present incentivises activities that will forestall future, more valuable uses to some degree.

In the Western United States, property rights in water streams are recognised in accordance with a prior appropriation doctrine. If one diverts water from the river onto one's land and makes use of it for irrigation, for example, one acquires a right to this flow. However, if one chooses to leave the water running adjacent to one's land alone simply to preserve it as an end unto itself or for its inherent beauty or for potential future use, one acquires no rights whatsoever. Resultantly the value one currently derives from the water being left in this way has no rights-based protection, neither do the option-values, meaning that someone could come along and divert water from that river and *they* would be protected in doing so because they are making ostensibly productive use of the water.[50]

However, if we look beyond the idea that uses must be *productive* in order to acquire ownership rights, and rather recognise that any use—broadly understood—could constitute an activity worthy of rights-based protection to the extent that befits the activity. We need not think of it as binary in the sense of you either take the water and own it, or you don't take the water and have no rights at all. If one fishes in the river, this constitutes a use of the river and should be granted legal protection; and that means placing limits on the degree to which the river can be diverted because doing so would affect the fisherman's prior use. Those who canoe down the river for pleasure should be protected in this activity just as anyone making use of it for agricultural purposes. There are all sorts of other uses which, whilst perhaps not always monetisable, are uses of the world's resources protection of which is required on the same grounds as using the water to irrigate commercial farms, and will have the result of preserving the river as it is to a significant extent. Often, it will be present use-values that are also destroyed alongside option-values, and if those present uses are given property protection, so may the option-values.

Where an ecosystem service is truly essential to the ongoing lives of the nearby population—to their ongoing activities and to their health, and so forth—it should be recognised that these people have acquired rights to the ongoing integrity of those ecosystem services, and therefore easements that would diminish them may not be acquired by new users.

In addition to this, we should consider the possibility that preserving marine life or mineral composition also constitutes use of a resource, in this case, the river, even if it does not involve constant physical intervention in it by a human hand. Indeed, it is widely recognised that leaving a field fallow for a season or a year counts as a use of the field—a use that can be interfered with by other people polluting the field or otherwise making use of it themselves. This is the case because letting a field lie fallow is a commonly understood activity in all agricultural communities. Let the active preservation of ecosystem services and natural resources also become a publicly intelligible activity. The more we do it, the more refined the rules that constitute it can become, and more refined the contours of the property protections required can be.

Conclusion

In this essay, I have attempted to outline what natural rights theory is, and how it illuminates the moral nature of the problem of pollution. In this view pollution is a problem much like any other that involves conflict over the use of scarce socioeconomic space. I have offered some typology for considering how problems of pollution emerge as a result of changing social attitudes, objective circumstances, and human knowledge. The concern that natural rights theories assign property protections only to economically productive uses of resources (including pollution easements) I have argued is a serious one. However, when we expand our lens of human activities deserving of rights-based protection beyond those which are ostensibly, immediately productive in monetisable ways, natural rights also grant protections to activities that involve holding back resources for future use, for actively preserving their ecological composition or ecosystem services. So long as activities can be recognised intelligibly by the public as activities, then they can ground property-based protection.

Notes

1. Jack Donnelly, "Human Rights as Natural Rights," *Human Rights Quarterly* 4, no. 3 (1982): 391–405; Finnis, John Finnis, "Grounding Human Rights in Natural Law," *American Journal of Jurisprudence* 60 (2012): 195–225; James Nickel, "Human Rights," in *The Stanford Encyclopedia of Philosophy*, ed. Edward N. Zalta (Stanford University Metaphysics Research Lab, 2019), https://plato.stanford.edu/archives/fall2021/entries/rights-human.
2. Heavy work is often made of the distinction between secular and theological derivations of natural rights, but these are simply disagreements about what nature is or where nature comes from, not about how rights are grounded in it.
3. If non-human animals have natural rights then these will also be natural moral facts about these creatures.
4. Sometimes some aspect of morality might be served by violating a right, but at least one aspect of morality always requires that rights be

respected: justice. It is never *just* to violate a rights, even if there may at times be other aspects of morality that recommend doing so.

5. It is by assigning spheres of permissible and non-compulsory choices that rights are said to make individuals into small-scale sovereigns; H. L. A. Hart, *Essays in Jurisprudence and Philosophy* (Oxford: Oxford University Press, 1985), 183.
6. Indeed, it is essential that all compulsory acts also be permissible—we must be within our rights when we act on duty; Hillel Steiner, "The Structure of a Set of Compossible Rights," *The Journal of Philosophy* 74, no. 12 (Dec. 1977): 767–775; Hillel Steiner, *An Essay on Rights* (Oxford: Blackwell, 1994), 86–101; Robert E. Goodin, "The Duty to Let Others Do Their Duty," *The Journal of Ethics* 24 (2020): 1–10, https://doi.org/10.1007/s10892-019-09318-x.
7. John Locke, *Two Treatises of Civil Government*, ed. Thomas Hollis (London: A. Millar, 1764), sec. II.VII–IX.
8. Immanuel Kant, *The Metaphysics of Morals*, ed. Mary Gregor (Cambridge: Cambridge University Press, 1991), pt. I.
9. One might also acquire rights through bequest, though this strictly speaking should be regarded as a form of transfer. Often it is argued that one can also acquire rights through restitution (getting them back after they have been wrongfully taken away), but this is mistaken. One's *rights themselves* can never be wrongfully taken away. They can be violated, but they cannot be removed. So, restitution is just third parties bringing rights violations to an end; Billy Christmas, *Property and Justice: A Liberal Theory of Natural Rights* (London: Routledge, 2021), 46; Judith Jarvis Thomson, "Rights and Compensation," in *Rights, Restitution, and Risk: Essays in Moral Theory*, ed. William Parent (Cambridge: Harvard University Press, 1986), 66–77; Jeremy Waldron, "Superseding Historic Injustice," *Ethics* 103, no. 1 (1992), 14.
10. Christmas, *Property and Justice: A Liberal Theory of Natural Rights*; Erick Mack, "Self-Ownership and the Right of Property," *The Monist* 79 (1990): 519–543.
11. Kant, *The Metaphysics of Morals*, sec. 6.243–6.250; Louis-Philippe Hodgson, "Kant on Property Rights and the State," *Kantian Review* 15, no. 1 (2010), 58–63; Author Ripstein, *Force and Freedom: Kant's Legal and Political Philosophy* (Cambridge: Harvard University Press, 2009), 95–96; Anna Stilz, *Liberal Loyalty: Freedom, Obligation, and the State* (Princeton: Princeton University Press, 2009), 42; Helga Varden, "Kant's

Non-Voluntarist Conception of Political Obligations: Why Justice Is Impossible in the State of Nature," *Kantian Review* 13 (2008): 6–8.

12. Robert Filmer, *Patriarcha: Of the Natural Power of Kings* (London: Richard Chiswell, 1680).
13. Natural rights theory therefore offers a *moralised* conception of liberty: you are free to act within your rights. Restrictions to your action that do not violate your rights do not restrict your liberty. Acting outside of your rights is not liberty, but licence; Ralf M. Bader, "Moralizing Liberty," in *Oxford Studies in Political Philosophy*, eds. David Sobel, Peter Vallentyne, and Steven Wall, vol. 4 (Oxford: Oxford University Press, 2018), 141–166; Ralf M. Bader, "Moralized Conceptions of Liberty," in *The Oxford Handbook of Freedom*, eds. David Schmidtz and Carmen E. Pavel (Oxford: Oxford University Press, 2018), 59–75; Christmas, *Property and Justice: A Liberal Theory of Natural Rights*, 37–43.
14. When the risk of a rights violation being imposed is significant enough, natural rights theorists widely agree that it constitutes a rights violation itself; Judith Jarvis Thomson, "Imposing Risks," in *Rights, Restitution, and Risk: Essays in Moral Theory*, ed. William Parent (Cambridge: Harvard University Press, 1986), 173–191.
15. Olle Torpman, "Libertarianism, Climate Change, and Individual Responsibility," *Res Publica* 28, no. 1 (2022): 125–148; Matt Zwolinski, "Libertarianism and Pollution," *Philosophy and Public Policy Quarterly* 32 (2014): 10–11.
16. If the restitution due is lower than the price that would be negotiated, then the polluter has an incentive to pollute without permission and pay the fine rather than the fee, since it is cheaper. In such a case, the rate of restitution is either set too low and should be increased, or enforcement is needed to stop the pollution. It should be noted that the latter is also not costless.
17. Jonathan Benson, "Environmental Law and the Limits of Markets," *Cambridge Journal of Economics* 42, no. 1 (2018): 215–30.
18. That purchase price would be reflective of the saving *A* would have made on her apartment, given that it did not come with an easement right to peace and quiet at night.
19. Murray N. Rothbard, "Law, Property Rights, and Air Pollution," in *The Logic of Action Two: Applications and Criticism from the Austrian School* (Cheltenham: Edward Elgar Publishing, 1997), 145–48.
20. Ronald H. Coase, "The Problem of Social Cost," *Journal of Law and Economics* 3 (1960): 1–44.

21. Richard A. Epstein, *Takings: Private Property and the Power of Eminent Domain* (Cambridge: Harvard University Press, 1985); Richard A. Epstein, "Property Rights, State of Nature Theory, and Environmental Protection," *New York University Journal of Law and Liberty* 4 (2009): 1–35.
22. Joel Feinberg makes the distinction between rights *violations* (that you have a right to repel) and rights *infringements* (that you have no right to repel but do have a right to compensation ex post): Joel Feinberg, "Voluntary Euthanasia and the Inalienable Right to Life," *Philosophy and Public Affairs* 7 (1978): 93–123: Drawing on the more general formulation from Judith Jarvis Thomson, "Some Ruminations on Rights," in *Rights, Restitution, and Risk: Essays in Moral Theory*, ed. William Parent (Cambridge: Harvard University Press, 1986), 49–65.
23. *Cf.* Daniel M. Hausman, "The Impossibility of Interpersonal Utility Comparisons," *Mind* 104, no. 415 (1995): 473–490.
24. Billy Christmas, "Incommensurability and Property Rights in the Natural Environment," *Environmental Politics* 26, no. 3 (2017), 502–520.
25. One of the fundamental grounds of rights is that we each have different values, and will conflict in our pursuit of those values in the context of scarce resources; Christmas, *Property and Justice: A Liberal Theory of Natural Rights*, 17–22.
26. Ibid., 119–26.
27. I have never seen a theory that claims to give weak, fragile, relative, and conditional protection to individuals from wrongful treatment, so I've never understood what plain of contrast such characterisations of natural rights are attempting to illuminate.
28. Peter Railton, "Locke, Stock, and Peril: Natural Property Right, Pollution and Risk," in *To Breathe Freely: Risk, Consent, and Air*, ed. Mary Gibson (Totowa: Rowman and Allanheld, 1985), 89–123.
29. Although these things are not irrelevant to the gravity of the rights violation, they are presumptively irrelevant to the fact of the rights violation.
30. David Friedman makes similar remarks; that if it is impermissible for my to fire a laser beam at your property because I didn't have consent to "use" your property, then it must also be impermissable if I lower the water of the laser beam to the point where it is just a beam of light. In both cases, I non-consensually make use of your property. David

Friedman, *The Machinery of Freedom: A Guide to Radical Capitalism*, 2nd ed. (LaSalle: Open Court Publishing Company, 1989), 168.

31. David Sobel, "Backing Away from Libertarian Self-Ownership," *Ethics* 123 (2012): 12.
32. Eric Mack, "Elbow Room for Rights," in *Oxford Studies in Political Philosophy*, eds. David Sobel, Peter Vallentyne, and Steven Wall, vol. 1 (Oxford: Oxford Online, 2015).
33. This approach is sometimes known as "specificationism." Cf. Russ Shafer-Landau, "Specifying Absolute Rights," *Arizona Law Review* 37, no. 1 (1995): 209–224; John Oberdiek, "Specifying Rights Out of Necessity," *Oxford Journal of Legal Studies* 28, no. 1 (2008): 127–146; Roderick T. Long, "In Praise of Bleeding Heart Absolutism," *Cato Unbound* (blog), 2012. https://www.cato-unbound.org/2012/04/04/roderick-t-long/praise-bleeding-heart-absolutism/.
34. Christmas, *Property and Justice: A Liberal Theory of Natural Rights*, chap. 3.
35. Hillel Steiner, "Debate: Levels of Ideality," *The Journal of Political Philosophy* 25, no. 3 (2017): 376–384.
36. Carmen Pavel argues the following.

    > The absolutist character of this understanding of [natural] property rights seems to preclude reasonable compromises. Yet assessing what counts as aggression, and what is the threshold above which pollution is damaging and must be curtailed, is why we need conventions in the first place. There is nothing else that could replace conventions in the first place.

    Carmen E. Pavel, "A Legal Conventionalist Approach to Pollution," *Law and Philosophy* 35 no. 4 (2016): 349.
37. John R. Searle, *Speech Acts: An Essay in the Philosophy of Language* (Cambridge: Cambridge University Press, 1969), 33–42; John R. Searle, *The Construction of Social Reality* (New York: Simon & Schuster, 1995): chap. 2.
38. Christmas, *Property and Justice: A Liberal Theory of Natural Rights*, 145–152.
39. As Wittgenstein puts it, "not agreement in opinions, but in form of life" Ludwig Wittgenstein, *Philosophical Investigations* translated by G. E. M. Anscombe, 3rd ed. (Oxford: Blackwell, 1986), sec. 241.
40. Peter Winch, *The Idea of a Social Science and Its Relation to Philosophy*, 2nd ed. (London: Routledge, 1990), 86–87.

41. David Hume, *A Treatise of Human Nature*, ed. Lewis Amherst Selby-Brigge, 3 vols. (Oxford: The Clarendon Press, 1739), sec. III.I–III.II; Carl Menger, *Principles of Economics* translated by James Dingwall and Bert F. Hoselitz (Auburn: The Ludwig von Mises Institute, 2007): chaps 1–2; Léon Walrus Walras, *Elements of Pure Economics: Or the Theory of Social Wealth* translated by William Jaffé (London: Routledge, 1954): secs 21, 23; Christmas, *Property and Justice: A Liberal Theory of Natural Rights*, chap. 1.
42. One of the observations made by Elinor Ostrom about the governance of common-pool resources is that often the nuances of disposition are more readily known—often tacitly know—by those who regularly use it, not least because it forces them to interact with other uses with the same regularity, and come to understand them Elinor Ostrom, *Governing the Commons: The Evolution of Institutions for Collective Action* (Cambridge: Cambridge University Press, 1990); There are similar themes regarding the importance of tacit knowledge and local understandings at work in the analysis from James C. Scott, *Seeing Like A State: How Certain Schemes to Improve the Human Condition Have Failed* (New Haven: Yale University Press, 1998), chap. 9.
43. Charles Taylor, "Interpretation and the Sciences of Man," *The Review of Metaphysics* 25, no. 1 (1971): 3–51.
44. Bamford v. Turnley, 122 27 (Court of the Exchequer 1862), reported by George Bramwell; Billy Christmas, "Answering the Conventionalist Challenge to Natural Rights Theory," *Res Publica* 27, no. 3 (2021), 343–345.
45. I am grateful to Aimée Hobbs for drawing my attention to this additional category of cases.
46. John Locke, *Two Treatises of Civil Government*, sec. II.V.27.
47. Robert Nozick, *Anarchy, State, and Utopia* (New York: Basic Books, 1974), 174.
48. Bas van der Vossen, "John Locke on Commercial Society," *Liberty Matters* (blog), 2021. https://oll.libertyfund.org/page/liberty-matters-john-locke-on-commercial-society-september-2021#Van%20der%20Vossen%20lead.
49. Cf. Brian Kogelmann, "Lockeans Against Labor Mixing," *Politics, Philosophy and Economics* 20, no. 3 (2021): 251–272.
50. Cf. Bas van der Vossen, "Property, the Environment, and the Lockean Proviso," *Economics and Philosophy* 37, no. 3 (2021): 395–412.

Do Libertarians Have Anything Useful to Contribute to Climate Change Policy?

Daniel H. Cole

Introduction

Libertarians have written a lot about environmental protection, including both natural resources conservation and pollution control. Some of them have even tackled the largest and most complex environmental problem humans have ever confronted: anthropogenic climate change. This chapter's purpose is to assess the practical utility of libertarian approaches to environmental and climate policies. It presumes the consensus scientific predicate supporting the goal of mitigating greenhouse gas emissions to avoid dangerous anthropogenic interference with the climate system.

No consensus set of libertarian environmental and climate policies exists. However, all libertarians agree that property-based solutions are the only truly legitimate approach, at least when they are available,

D. H. Cole (✉)
Indiana University Maurer School of Law, Bloomington, IN, USA
e-mail: dancole@indiana.edu

J. H. Adler (ed.), *Climate Liberalism*, Palgrave Studies in Classical Liberalism,
https://doi.org/10.1007/978-3-031-21108-9_3

because they are non-coercive, and they are believed to be both necessary and sufficient to deal with environmental problems. Beyond that, disagreements are rife. Some libertarians argue that property-based solutions are always available. Others concede that property solutions are sometimes unavailable or too costly, but nevertheless deny any legitimate role for government intervention. Still others admit an occasional need for government intervention but differ on exactly what regulatory policies are legitimate.

The first section of this chapter on "The Varieties of Libertarian Environmental and Climate Policies" outlines and attempts to organize a wide array of libertarian environmental and climate positions in relation to one another. The next section on "Special Problems of Climate Change for Free Market Environmentalism" assesses the theoretical and empirical plausibility and utility of various libertarian environmental and climate policies. The chapter concludes that even the most generous libertarian policy proposals are insufficient to grapple with the complexities of climate change and recommends that libertarians need to (a) move beyond generalized policy positions and focus more on the design and implementation of specific policies in real-world contexts, and (b) adopt a more pragmatic, problem-solving approach to climate change that understands policies as experiments, in the Popperian tradition.

The Varieties of Libertarian Environmental and Climate Policies

Libertarian Dissensus

Libertarianism is one school of thought, among several others, in the classical-liberal tradition.[1] What sets libertarianism apart from its siblings is its laser-like focus on individual liberty and property rights. While other schools of liberal thought are also committed to individual liberty and property, they are not treated as exclusive maximands. For instance, conservative-liberals, following Burke and Hume, are committed to social, cultural, political, and religious traditions that are not always consistent with individual autonomy. Moderate liberals, like Mill and

Popper, and progressives, like Dewey, also espouse commitments to political equality, which sometimes require higher levels of government interference in markets than libertarians would countenance, e.g., to prevent unlawful discrimination.

The libertarian literature describes a surprising lack of consensus about what libertarianism requires. All libertarians share a strong preference for private ordering, including for resolving environmental problems. But some libertarians insist that property-based solutions, alone, are legitimate, while others are willing to countenance various margins of government intervention. Even when libertarian thinkers reach similar policy conclusions, including on problems like climate change, they may do so for widely differing—sometimes, mutually inconsistent—reasons. For example, the respective environmental policy conclusions of Murray Rothbard's "Law, Property Rights, and Air Pollution"[2] and the first edition of Terry Anderson and Donald Leal's *Free Market Environmentalism*[3] are nearly identical, even though Rothbard derives his position from a moral theory premised on natural law, while FME is based on welfare-consequentialism. At other times, libertarian thinkers, starting from a shared set of principles, come to very different policy conclusions. For example, among libertarians who agree with the need for some government regulation to protect the natural environment or the global climate, some consider property-like mechanisms, such as cap-and-trade schemes, to be the only legitimate form of regulation,[4] while others consider such schemes wholly inappropriate on libertarian grounds.[5] Similar disagreements arise with respect to effluent taxes.

This libertarian dissensus explains why any attempt to identify *the* libertarian climate policy would be futile. The best we can do is to outline a few distinct libertarian environmental/climate policy positions (Table 1).

Table 1 Libertarian climate policy positions

I. Property-based
A. Property-based whenever feasible
B. Property-based only
II. Property-based plus limited government intervention
A. Non-regulatory interventions only
B. Regulatory interventions limited to "market-based" instruments
1. Effluent taxes only
2. Cap-and-trade only
3. Either taxes or cap-and-trade or both
III. Regulatory intervention with a preference for "market-based" instruments, but not ruling out other instruments

Libertarian Climate Policies

Pure Property-Based Environmental and Climate Policies

As noted, the only consensus among libertarian environmentalists is a unanimous preference for property-based solutions, i.e., enforcement of property rights in common-law private tort actions, when they are feasible. Private lawsuits in common-law courts to protect property from harm does not, after all, involve government coercion (assuming the judicial process itself is unbiased). But the unanimity quickly evaporates with respect to problems that cannot be effectively resolved by common-law courts. Some libertarians insist that the property rights approach is the only legitimate approach, period,[6] while others allow for various margins of government intervention when property-based solutions are not practicable.[7]

Another libertarian dissensus exists with respect to the appropriate remedy in pollution-related common-law tort cases. The default libertarian position is that the appropriate remedy for a harmful property invasion is injunctive relief to stop the invasion.[8] Money damages cannot substitute for injunctive relief because that would, in effect, amount to a forced transfer of the plaintiff's property to the defendant at a price set by the court. As Jules Coleman and Jody Kraus explain,[9] "[i]t is surely odd to claim that an individual's right is protected when another individual is permitted to force a transfer at a price set by third parties. Isn't the very idea of a forced transfer contrary to the autonomy or liberty thought constitutive of rights?" However, injunctions can have large negative impacts on social welfare, especially when the pollution source provides net social benefits. In addition, injunctive relief sometimes provides no remedy at all, e.g., for Pacific Islanders whose homes are submerging because of climate change-related sea-level rise. This and similar cost-related issues motivated Robert Nozick to adopt a "Principle of Compensation,"[10] according to which money damages could substitute for injunctive relief if they are sufficient to make the plaintiff indifferent between suffering no invasion or suffering the invasion and receiving the compensation. Note the substantial trade-off involved. Nozick's Compensation Principle avoids the potentially massive economic and social welfare consequences of injunctions that could bring economic activity to a virtual halt, but only at the price of legitimating forced judicial transfers of property.

The dominant theory of libertarian environmental policy is Free Market Environmentalism (FME). In their 1991 book of that title, Anderson and Leal observe that some environmental goods are more difficult and costly to privatize than others.[11] They concede that defining, allocating, and enforcing property in the atmosphere, for example, presents great challenges.[12] But they consider this a technological problem and anticipate the innovation of "new forms of fencing" technologies that would reduce the costs of creating atmospheric boundaries, just as the invention of barbed wire in the 1870s facilitated settlement of the western US by reducing the costs of enforcing land boundaries.[13] In the meantime, they recommend doing nothing (at least, that was their recommendation before 2001).

FME has few specific recommendations for climate policy. In 1991, Anderson and Leal recommend ending subsidies for beachfront construction, relocating agricultural activities as necessitated by changing patterns of precipitation, and developing more drought-resistant crops.[14] "In the case of automobile emissions, which are a major source of greenhouse gases," they observe, "the property rights approach is quite simple: privatize congested highways."[15] "For other sources of pollution where the property rights solution is not as simple, more reliance must be placed on the evolution of property rights."[16] In other words, in the absence of a property-based solution, just wait for a property-based solution.

Initially, FME opposed property-*like* regulatory measures, such as cap-and-trade, because "[a] government agency must still determine the level of permits, and the permits do not force polluters to compensate those harmed by the pollution. In this system the political process determines the initial or optimal pollution levels, not the polluters bargaining with those who bear the costs of the pollution."[17] However, the FME position has evolved over time. By 1998, Anderson and Leal were prepared to concede "a utilitarian argument for considering government intervention," though they did not provide an example of an intervention FME might condone.[18] That came later, in 2001, in a revised edition of their 1991 book. The revised FME fully endorses tradeable permits for pollution control.[19]

Property-Based Plus Limited Government Action

Not all "libertarian environmentalists" were satisfied with FME, which in its original version declared that private property lawsuits are the only legitimate means of resolving pollution problems. The economist Milton Friedman, who considered himself "a libertarian with a small l,"[20] followed the more conventional welfare-economics approach, treating pollution as a negative externality—a market failure that warrants government intervention.[21] Other libertarian scholars, such as Jonathan Adler,[22] have supported non-regulatory interventions as substitutes for property-based solutions that are not available. Still others, such as Edwin Dolan,[23] support tradeable permit schemes, a property-*like* form of regulation, *or* pollution taxes, but rarely both.[24] It is hard to find a libertarian environmentalist who embraces both carbon taxation and tradeable permitting schemes. Even more rare is a libertarian environmentalist who would even consider traditional forms of regulation (a.k.a., command-and-control).

In 2009, the libertarian legal scholar Jonathan Adler came out in support of non-regulatory government initiatives, including "cash payments, development assistance, funding of adaptive measures to increase overall resilience and reduce climate-vulnerabilities" in less-developed countries. These payments would be a substitute for common-law remedies that simply do not exist.[25] "Such compensation would not have to take the form of government-to-government transfers, however."[26] Individual emissions sources could "fund various adaptive measures directly."[27] Perhaps in recognition of that heroic assumption, Adler does not profess a great deal of faith in these kinds of measures, but he considers them far better than doing nothing.

What Adler did not support, at that time, was any kind of government regulation of emissions:

> [c]entralized federal regulation is a poor match for the climate policy challenge. Regulatory requirements may work tolerably well at mandating the diffusion and adoption of viable technologies, but command-and-control regulation has a poor record of driving technological advancement in a desired direction, particularly where revolutionary innovations are required. Market-based regulatory systems, including cap-and-trade systems, have not fared much better.[28]

This does not mean that Adler believed that common-law tort suits are necessarily preferable to regulatory solutions. He expressly notes that, "even though regulation is not necessarily effective, it may be more effective than the common law."[29]

More recently, however, Adler has come out in favor of carbon taxation because, compared to cap-and-trade schemes, it is easier to design, enact, and administer, more difficult to challenge legally, and can be made revenue neutral by replacing taxes on goods, such as income.[30] Arguably, a revenue-neutral carbon tax, i.e., a carbon tax that is offset by reductions in taxes on goods such as income or sales, is consistent with fundamental libertarian commitments to liberty and property because it would not increase government coercion on net. At the same time, it would deal with the harms caused to individuals and their property by greenhouse gases.

Tax policy has long divided libertarians. Rothbard famously wrote that "taxation is robbery."[31] Nozick, on the other hand, recognized that even a minimal state would require some amount of tax-based funding.[32] He established his Principle of Compensation as part of his "nightwatchman state" to allow for the internalization of externalities via money damages as a legitimate remedy for unjustified invasions of property rights. Adler's position, like Nozick's, is based on the transaction costs of defining, allocating, and enforcing property rights in common-law courts. It amounts to tacit recognition that we are in a second-best world, not the world of costless transacting assumed by neoclassical economic theory and purist libertarians like Rothbard.

Contra Adler, libertarian economists Jeffrey Miron and Pedro Braga Soares oppose carbon taxes.[33] "The first problem is setting the tax at the right level."[34] Set too high, the tax will lead to inefficiently high cuts in emissions; set too low, it will fail to cut emissions enough. Getting the tax level right requires an accurate estimation of the social cost of carbon, which is subject to a lot of uncertainty. Differing initial assumptions lead to widely varying estimates, which are also extremely sensitive to parameter values for the discount factor. Taken together, these problems amount to an instantiation of Hayek's "knowledge problem."[35] Consequently, Miron and Soares default to the original FME position on climate change: "even if anthropogenic climate change will impose significant costs, the least bad policy response might be doing nothing other than focusing policy on accelerating economic growth."[36] The argument boils down to this: since a carbon tax might fail, we should act as if the social cost of carbon is zero.

Other libertarians, such as Edwin Dolan,[37] prefer tradeable permits over effluent taxes to deal with pollution problems when privatization is not feasible. Still others staunchly oppose tradeable permits based on the same arguments Anderson and Leal made in 1991: some government agency must set and allocate the initial quota limits and those who suffer harm from the pollution are not compensated.[38] Robert McGee and Walter Block denigrated cap-and-trade schemes as a form of "market socialism."[39]

More difficult to place among libertarian environmentalists is Martin Anderson, who served in the Reagan White House. In a brief and cryptic piece of puffery in support of George H.W. Bush's unsuccessful 1990 presidential campaign, Anderson argues that neither cap-and-trade nor environmental taxes are legitimate because they do not sufficiently protect the property rights of those who are harmed by pollution.[40] Charging a fee for pollution "is like taxing burglars as an economic incentive to deter people from stealing your property, and just as unconscionable.... *What we need are tougher clearer environmental laws that are enforced – not with economic incentives but with jail terms.*"[41] The reference to "clearer environmental laws" is intriguing. Is he promoting government regulation of pollution? It is not entirely clear. On the one hand, Anderson cites with apparent approval President George H.W. Bush's support for a complete ban on ocean dumping of sewage sludge (a classic piece of command-and-control regulation), speeding up remediation of hazardous waste sites under federal regulatory laws, and federal government leadership on protecting groundwater supplies. On the other hand, he writes at the very end of the article, "President Bush and the Republican Party have a historic opportunity to reclaim the environmental heritage of Teddy Roosevelt. All they need to do is reaffirm the ancient principle of private property rights and devise innovative ways to implement those rights in a modern technological society."[42]

Perhaps the most generous libertarian environmental policy, *if it is libertarian*, comes from the libertarian legal scholar Richard Epstein.[43] In a paper nominally about the 1980 US Superfund Law,[44] Epstein articulates a cogent set of principled and pragmatic propositions[45]:

1. Principles of justice that protect liberty and property often require government action for the protection of the environment;
2. These principles do not determine the precise form of remedial measures, nor whether the public or private sector should initiate them;
3. The selection of the proper remedy involves difficult empirical questions that must be resolved by some utilitarian calculus; and
4. The major and lasting criticism to be made of government environmental policy is not that the state has moved in areas unworthy of governmental intervention, but that its chosen modes of intervention are ill-calculated to achieve their objectives.

Proposition (3) implies different remedies might be appropriate in differing circumstances. Epstein, who reveres common-law history as much as any US legal scholar, nevertheless doubts its utility for dealing with large-scale pollution problems. He points to the standard list of evidentiary problems that plague environmental plaintiffs, which other libertarians either do not take seriously enough or elide altogether.[46]

Evidentiary problems include identifying sources responsible for harmful pollution, which might be located hundreds of miles from where the harm occurs. Even when the relevant pollution sources are local, it might be impossible to pick out the responsible polluter when the harm is caused by pollution from a vast number of sources, and no single source emits enough of the pollutant to cause the harm. An example would be the urban smog produced by hundreds of thousands of automobiles in a metropolitan area. On the other hand, a major emitting source might escape liability for causing widespread harm because the harm they cause to any single individual or property is too slight to justify the cost of filing suit. In addition, long lag-times between exposure to a pollutant and manifestation of harm, which is especially common in the case of toxic/carcinogenic pollutants creates significant problems relating to the preservation of evidence over the course of decades.[47]

Because of these various problems, Epstein explains,[48] "[t]he conditions of environmental cases therefore call for a specific examination of private liability rules and government regulation as substitutes." Though highly critical of the Superfund statute in general, Epstein expressly supports a "more modest" government regulatory policy on hazardous wastes based on three elements[49]:

1. The government should be allowed to maintain an action for the destruction or damage of unowned resources. This is designed to prevent destruction of the common pool, but requires no special powers of taxation…;
2. The government should operate a permit system that allows it to identify the individual defendant responsible for the release of a chemical substance into the environment…; and
3. The government should be given broad and immediate powers to clean up and regulate existing dumps. They present the greatest threats of large-scale pollution and the greatest obstacles to private relief, given the weaknesses of private damage actions after the fact,

> and the near impossibility of organizing a coalition of plaintiffs to obtain injunctive relief beforehand. Concentration on these sites makes it inappropriate to worry about the tax on inputs; it would be better to devise a system of fines and charges that could be levied against the owners or managers of the plants in question, or the parties who have abandoned them. There will be some instances in which these parties cannot be identified, and to fund those necessary cleanup operations it may be necessary to draw upon general revenues.

Not surprisingly, libertarian commentators chided Epstein for staking out a position that was insufficiently libertarian.[50] This raises a serious question about the boundaries of libertarianism as a distinctive school of classical-liberal thought. The fact that hardly any libertarian scholars would support Epstein's pragmatic embrace of government regulation in circumstances where property rights could not be effectively defined and enforced highlights the myopic focus of libertarian theory on private property, contracts, and markets as the only legitimate institutions of social coordination.

Epstein may be an outlier among libertarian scholars, but he is not entirely alone. Peter Hill declared a similar willingness to at least consider some kinds of government regulations for climate policy[51]:

> Major problems of global climactic change and of pollution that crosses political boundaries are much more difficult to solve. Taxes, subsidies, tradeable pollution permits, and treaties among political units may be appropriate to approximate markets and property rights.

Hill nevertheless appreciates that "[a]ttempts to use government to stimulate the solutions of the marketplace will be plagued by both information and incentive problems. Thus, government ought to move with extreme caution in implementing solutions of this type."[52]

Special Problems of Climate Change for Free Market Environmentalism

The practical difficulties of libertarian climate policy include: (1) the global nature of the climate problem, which creates collective-action problems at the international level as well as the national, state, and local levels of governance; (2) intractable impediments to international, as well as domestic, climate tort litigation; and (3) the dubious track record of regulatory tools some libertarians prefer, such as cap-and-trade. This is not to say that private-ordering solutions have no role to play in mitigating climate change. Private organizations like the World Business Council for Sustainable Development have already demonstrated an ability to obtain voluntary emissions reductions from greenhouse gas emitters.[53] Nevertheless, the idea that the problem of avoiding dangerous anthropogenic interference with the climate system might be resolved without any coercive government measures is wishful thinking at its worst.

Problems for Common-Law Property-Based Solutions

Libertarians must be aware that their preoccupation with American-style property and the American common law ignores the majority of countries around the world with very different property systems and no common law. Even among countries that have common law, rules differ substantially. For example, the US recognizes regulatory takings as a legal cause of action; the UK does not.[54] Harmonizing the common law across the UK and US would be highly improbable, let alone extending harmonized common-law rules to every country on earth. The notion of a global property rights regime is fanciful; if it were not, libertarians might well oppose it as coercive.

As Adler has noted,[55] the international community does have a kind of common law, known as customary international law. He explains that "[c]ustomary principles of international law recognize that nations are responsible for polluting activities conducted within their borders that harm people or property in other nations."[56] But his further claim that "there is no international tribunal capable of adjudicating such claims in the context of global warming" is technically incorrect.[57] In fact, the International Court of Justice (ICJ) is fully capable of adjudicating claims based on customary international law, including for harms from climate change. It even provides for a form of injunctive relief, known as "provisional measures of protection,"[58] as well as compensatory damages. However, it is difficult to imagine the ICJ enjoining a defendant country from emitting greenhouse gases.

In any case, the utility of the ICJ is decidedly limited. It suffers from an immense procedural impediment: it cannot hear a case unless the defendant voluntarily consents to its jurisdiction. It is difficult to imagine any member state of the United Nations opening itself to climate change liability in this way. In addition, the existence of several climate change-related treaties, including the United Nations Framework Convention on Climate Change, the Kyoto Protocol, the Paris Agreement, and even the Ozone Accords, arguably disables customary international law as a basis for litigation.[59]

Another, albeit insufficient, adjudication option for parties suffering climate harm, such as small Pacific Island nations, would be to seek a ruling from the UN General Assembly, sitting as a trier of fact. The assembly would deliberate over the evidence and declare whether or not a major emitting country is responsible for harm caused to Tuvalu or Bangladesh. But the UN General Assembly has no authority to offer a remedy.[60] And, of course, any "adjudication" in the General Assembly would be political, rather than legal.

Aside from the fact that only a plurality of the world has common law, there is no evidence that common-law tort lawsuits might avail potential plaintiffs in US courts. So far, more than two dozen common-law tort suits have been filed in US courts seeking damages and/or injunctive relief.[61] Every single one has failed at the first or second hurdle, as courts have invoked the Political Question Doctrine and Federal Common Law Preemption before ever reaching the merits.[62]

The strongest claim filed to date was *Native Village of Kivalina v. ExxonMobil* et al.[63] Situated at the tip of a long barrier reef extending from Alaska's northern coast, the Village of Kivalina was incorporated in 1969. Rising sea levels have been steadily eroding the rocks on which the

Village is perched. The Village can no longer be reached by road from the mainland. In fact, the barrier reef has been completely inundated, and the village had to be relocated in its entirety, at an estimated cost of $150–200 million.[64] In court, the Village alleged that its relocation was necessitated by rising sea levels, which were a direct consequence of climate change, caused in part by the defendants, including ExxonMobil. The Ninth Circuit never reached the merits of the case, concluding that its federal common-law jurisdiction was preempted by the US Clean Air Act[65] because of the US Supreme Court's decision in *Massachusetts v. EPA*,[66] which held that GHGs *could* be regulated under that statute.

Had its common-law jurisdiction not been preempted in the case, the Ninth Circuit in all likelihood would have dismissed the case anyway under the Political Question Doctrine, a prudential rule US courts use to avoid deciding cases based on current topics of serious debate among the political branches of government. Even if the case had survived all motions to dismiss, it is overwhelmingly likely the plaintiff would have lost on the merits due to causation-proof problems, including whether the Village had identified the right defendants, and whether the sea-level rise that caused the erosion could be fairly traced to emissions from the defendant's operations. Under a preponderance of the evidence standard, it seems highly unlikely that the defendants could have been found liable.

Supposing counterfactually that the defendants had been found liable in *Kivalina*, how much of an effect would it have had? It might have created incentives for other plaintiffs to sue American emitters of GHGs, although it is doubtful that plaintiffs outside the US would have any standing to sue in US courts. It is important to bear in mind, however, that a ruling from the Ninth Circuit is precedent only in the Ninth Circuit (including the lower federal district courts in the circuit). Even a favorable Ninth Circuit ruling would not necessarily have opened the floodgates to lawsuits in any other federal or state jurisdictions, let alone in other common-law countries. And no matter how much American courts or those in other common-law countries found GHG emitters liable for climate harm, it would have had no effect on GHG emitters anywhere else in the world. Given the challenges posed by climate change, even a successful litigation-focused strategy is only a band-aid.

Around the world, the courts have played practically no role in climate policy, with the singular exception of a court in the Netherlands, which ordered the government to enhance its response to climate change—hardly the kind of ruling libertarians have in mind.[67] On the other hand, some aspects of the current global system for controlling GHG emissions might, at first blush, appeal to some libertarians. Under the 2015 Paris Agreement,[68] each country voluntarily establishes its own "Nationally Determined Contribution" (NDC). But, of course, only some emission reductions will come from governments themselves.

Problems for "Market-Based" Regulatory Instruments

"Voluntary" or not, government commitments under the Paris Agreement will be enforced by national governments employing coercive measures against private enterprises. Around the world, we find in use the entire portfolio of regulatory mechanisms, including command-and-control mechanisms, emission taxes, carbon-content taxes on producer and/or consumer goods, cap-and-trade (i.e., performance-based standards with tradeable permits), various kinds of offset programs, and subsidies. Some libertarians insist on getting rid of all such regulatory instruments as a prelude to introducing a universal property-based solution.[69] No libertarian environmental policy examined in the previous section would embrace all the instruments in current use. Even Richard Epstein would likely object to command-and-control regulations, though he has not said so explicitly. While supporting government regulation of existing waste sites, he expressly refers only to fees, fines, and regulations requiring waste site managers to supply information.[70] And there is no basis for inferring that Epstein believes any form of public regulation would be either efficient or very effective.

The global climate agreements neither mandate nor rule out a particular approach to controlling GHG emissions, though they explicitly endorse tradeable permits and closely related offset programs, which many economists and some libertarian environmentalists prefer. The widespread support for cap-and-trade stems from the highly successful—economically, as well as environmentally—US acid rain trading program under the 1990 Clean Air Act Amendments.[71] In some respects, however, the lessons taken from that experiment were under-determined by the evidence. The acid rain program dealt with only two pollutants from a single type of industrial source, power plants, which were required to install expensive continuous emission monitoring (CEM) systems that had only recently become available. That equipment ensured accurate measurement of point source emissions without which no trading market for emissions permits would have gotten off the ground.[72] In addition, the success of the acid rain trading program depended on underlying institutional structures, including the rule of law, a non-corrupt bureaucracy to oversee the market, and a well-functioning market economy, one or more of which are missing in many countries around the world.[73]

In countries where budget and legal constraints on industrial emitters are soft, and where political interference and bureaucratic corruption are rife, such as China, no reason exists to believe that any kind of "market-based" mechanism for pollution control will be either efficient or effective.[74] In the 1970s, communist Poland began to implement the world's largest system of pollution taxes. It failed to reduce emissions at all, and not only because the tax level was too low. Pollution taxes, as well as other fees and fines, were treated as ordinary business expenses for which polluters were compensated in the following year's budget. Money removed from one pocket was replaced in the other. As it turned out, the only effective environmental protection measure in People's Poland was the most severe form of command-and-control: the government closed down some major pollution sources.[75] This hardly means that pollution taxes cannot work anywhere. British Columbia has, by all accounts, a successful carbon tax.[76] Could it work as well elsewhere? Almost certainly. Could it work as well everywhere? Certainly not. The same is true of emission trading schemes and offset programs.

Offset programs are a kind of emissions trading scheme. The 1997 Kyoto Protocol to the United Nations Framework Convention on Climate Change adopted two offset programs that are still in use[77]: Joint Implementation (JI), for transfers between two developed countries, and the Clean Development Mechanism (CDM), for transfers between a developed country and a developing country. In theory, both programs made sense, especially given the globally diffuse nature of GHG emissions. In practice, they have been plagued by problems.

The CDM program is more widely utilized than JI, but it would be a stretch to call it successful. Unlike the US acid rain trading program, CDM trades involve the entire portfolio of greenhouse gases and all industries, without any consideration of relative measurement and monitoring costs. In order for a source to receive credits toward meeting its individual emissions-reduction quota, the CDM project it sponsors must lead to verifiable emission reductions in the host country, and those reductions must be *additional* to reductions that would have occurred in the host country without the project.[78] This additionality requirement is critical because, without it, the only emissions reductions CDM projects generate would be on paper. The offsets would amount only to monetary transfers in exchange for credits, with no effects on actual emissions. In the climate regime, this is known as the problem of "hot air." The label applies not only to non-additional CDM credits, but also to excessive quota limits established for some countries, particularly East European countries in transition from communism. Kyoto Protocol quotas were percentage-based reductions from 1990 emission levels.

GHG emission levels in all of the former communist countries were much higher in 1990, the first year of their transitions from communism, than in subsequent years, as the move to competitive markets and hard budget constraints created incentives to improve dynamic efficiency in production, i.e., maximum output per unit of input. Consequently, they all received credits under the KP that greatly exceeded their actual emissions. These "hot air" credits could be sold to sources in other countries with substantial emission-reduction requirements. They were called "hot air" credits because any such sales entailed no actual emission reductions.[79]

The problems of additionality and hot air in the CDM program resulted in a great deal of fraud, mainly by China. A majority of all CDM projects have been hosted in China, rather than much-poorer developing economies that could have benefitted far more from the investments. And the vast majority of CDM projects approved in China involved have involved a single GHG, HFC-23, which is emitted in the production of the refrigerant HCFC-22. China announced plans to build more and more HCFC-22 plants, totally unrelated to economic demand, simply to facilitate lucrative CDM transfers that would curtail construction and operation of new facilities. In effect, the pollutant became the product. China's refusal, at that time, to share information on actual emission rates of HFC-23 or other GHGs made assessments of additionality especially difficult. Subsequent studies indicated that two-thirds or more of those projects led to the distribution of emission-reduction credits without any actual emission reductions.[80] The EU ended the charade in 2013, when it banned its member states from participating in CDM projects for HFC-23 emission reductions in China (European Commission, Directorate-General for Climate Action 2011).[81] Later that year, China entered into a bilateral agreement with the US to phase out HFC-22 emissions, which no longer possessed significant value.[82]

The moral of the story is that it takes a lot more than specifying quota limits and allowing trading to create a functional pollution market. Policymakers need to be careful about the institutional design and application of any specific mechanism for pollution control. The devil truly is in the details. But, so far, libertarian scholars have offered only broad policy recommendations favoring carbon taxes or tradeable permits, paying little or no attention to contextual factors or the specific institutional structures of those instruments. They have been better at explaining why any given policy cannot work in general, rather than explaining why a specifically structure policy might work better than others in a specific set of circumstances.

Implications and Conclusion

Abstract, simplistic, and ideal political theories are not particularly helpful in diagnosing and resolving complex collective-action problems in the real world. If libertarians are serious about contributing to climate policy debates, they need to do more than offer broad policy advice based on presuppositions. They need to become more pragmatic and dig into the details of institutional design and application, with an understanding that no single policy instrument (or set of policy instruments) is universally preferable in all circumstances.

This does not mean libertarians need to give up their strong commitments to preserving individual liberty and property; those are important considerations in policy discussions, as are practicable private-ordering solutions. But libertarians rule themselves out of serious political discussions when they continue to make claims like this one by Miron and Soares:

> [E]ven if anthropogenic climate change will impose significant costs, the least bad policy might be doing nothing other than focusing policy on accelerating economic growth. This maximizes the amount of resources available to adapt to climate change.[83]

The implicit and unfounded assumption that the US (and, by implication, the rest of the world) might simply grow its way out of the climate change problem is a clear indication that Miron and Soares take seriously neither the science of climate change nor the high risks associated with climate tipping points. Analytically, they are engaging in what Harold Demsetz called "the nirvana approach" to institutional analysis[84]: comparing an imperfect, real-world institutional arrangement (a carbon tax) and an ideal norm (*laissez-faire*).

For too long, libertarians have employed the "rhetoric of reaction" to reject every potentially useful regulatory policy, claiming they would have perverse consequences, be futile, or jeopardize liberty.[85] For Miron and Soares, the mere possibility that a carbon tax might get the prices wrong, means we should not even try it, never mind that even a suboptimal carbon tax could (1) reduce emissions efficiently and (b) the tax level could be adjusted to improve its effectiveness and efficiency. One thing we know with certainty, according to the Ninth Circuit U.S. Court of Appeals, is that: "the value of carbon emissions reduction is certainly not zero."[86] Even if we do not know precisely the social cost of carbon to determine the optimal tax, we might fairly suppose that we can set a tax rate somewhere between zero and the social cost level; and that even an imperfect tax, i.e., a tax that is not maximally efficient, might still yield net social benefits.

Yet, according to Miron and Soares, we should pretend the social cost of carbon is approximately zero simply because a carbon tax is costly to design and enforce, and its initial value might be suboptimally high or low. Doing nothing is sometimes sound policy advice, e.g., where the cost of doing anything is likely to exceed the cost of doing nothing.[87] But that certainly is not the case for climate change and many other environmental problems. It is entirely appropriate to argue about *how much* should be done, e.g., how much the US should invest in "climate insurance,"[88] but no good reason exists for pretending that the answer is to do nothing except maximize economic growth (using measures that ignore environmental costs).

Ultimately, every policy choice, including doing nothing, is an experiment.[89] Every experiment involves costs and risks, and many experiments fail, including do-nothing solutions. But we learn even from failed policy experiments. One lesson we should have learned a long time ago is that there are no panacea solutions to complex problems in the real world. As Elinor Ostrom has written[90]:

> Continuing to presume that complex policy problems are simple problems that can be solved through the adoption of simple designs that are given general names, such as private property, government ownership, or community organization, is a dangerous academic approach. Dichotomizing the institutional world into "the market" as contrasted to "the state" is so grossly inadequate and barren that it is surprising how the dichotomy survives as a basic way of organizing academic studies and policy advice. Oversimplification of our design options is dangerous since it hides more of the working parts needed to design effective, sustainable institutions than it reveals. And it reduces our awareness of the need to monitor outcomes and improve them over time through better processes of learning and adaptation.

Meanwhile, a "wealth of empirical information on real-existing property systems ... belies naïve and simplistic theories of property rights that reduce all resource conservation problems to either too little private-individual ownership or too little public ownership."[91] Instead of simplistic and inaccurate models of property systems, we require "theories that better account for the complexities and contingencies of actual resource governance regimes…."[92] Scholars, including libertarians, need to embrace complexity.

Notes

1. Daniel H. Cole and Aurelian Craiutu, "The Many Deaths of Capitalism," *AEON*, June 28, 2018, https://aeon.co/essays/reports-of-the-demise-of-liberalism-are-greatly-exaggerated.
2. Murray Rothbard, "Law, Property Rights, and Air Pollution," *Cato Journal* 2, no. 1 (1982): 55–99.
3. Terry Anderson and Donald Leal, *Free Market Environmentalism* (Boulder, CO: Westview Press, 1991).
4. Terry Anderson and Donald Leal, *Free Market Environmentalism*, Revised ed. (New York: Palgrave, 2001).
5. Martin Anderson, "George Bush Environmentalist," *The Christian Science Monitor*, January 4, 1989.
6. Rothbard, "Law, Politics, and Air Pollution;" Anderson and Leal, *Free Market Environmentalism* (1991).
7. *See* Jonathan H. Adler, "The Legal and Administrative Risks of Climate Regulation," *Environmental Law Reporter* 51, no. 6 (2021): 10485–10497; Epstein, Richard A. "The Principles of Environmental Protection: The Case of Superfund," *Cato Journal* 2, no. 1 (1982); Anderson, "George Bush Environmentalist;" Anderson and Leal, *Free Market Environmentalism*, Revised Edition (2001).
8. Murray Rothbard, *For a New Liberty: The Libertarian Manifesto*, 2nd ed. (Auburn: Ludwig von Mises Institute, 1973), 322.
9. Jules S. Coleman and Jody Kraus, "Rethinking the Theory of Legal Rights," *Yale Law Journal* 95, no. 7 (1986): 1338–1339.
10. Robert Nozick, *Anarchy, State, and Utopia* (New York: Basic Books, 1974), 78.
11. Anderson and Leal, *Free Market Environmentalism* (1991), 155.
12. Ibid.
13. Terry Anderson and Peter J. Hill. "The Evolution of Property Rights: A Study of the American West," *Journal of Law and Economics* 18, no. 1 (1973).

14. Anderson and Leal, *Free Market Environmentalism* (1991), 163–165.
15. Ibid., 165.
16. Ibid.
17. Ibid., 158.
18. Terry Anderson and Donald Leal, "Free Market Environmentalism: Hindsight and Foresight," *Cornell Journal of Law & Public Policy* 8, no. 1 (1998): 130.
19. Anderson and Leal, *Free Market Environmentalism*, Revised ed. (2001), 130–132.
20. Brian Doherty, "Best of Both Worlds: An Interview with Milton Friedman," *Reason Magazine*, June 27, 1995.
21. Milton Friedman, *Capitalism and Freedom* (Chicago: University of Chicago Press, 1962), 32–33; Milton Friedman and Rose Friedman, *Free to Choose* (New York: Harcourt, Brace Jovanovich, 1980), 213–218. I refer to Friedman as "quasi-libertarian" because, while he advocated for individual freedom, property rights, and small government, he never fit neatly into the libertarian family. Murray Rothbard referred to Friedman as "the Establishment's Court Libertarian" and a "statist;" Murray Rothbard, "Milton Friedman Unraveled," *Journal of Libertarian Studies* 16, no. 4 (2002): 37, 54.
22. Jonathan H. Adler, "Taking Property Rights Seriously: The Case of Climate Change," *Social Philosophy & Policy* 26, no. 2 (2009).
23. Edwin G. Dolan, "Controlling Acid Rain," in *Economics and Environment: A Reconciliation*, ed. Walter E. Block (Vancouver: Fraser Institute, 1990): 215–232.
24. See, e.g., Adler, "Legal and Administrative Risks"; Friedman and Friedman, *Free to Choose,* 217–18.
25. Adler, "Taking Rights Seriously: The Case of Climate Change," 315.
26. Ibid.
27. Ibid.
28. Jonathan Adler, "Heat Expands All Things: The Proliferation of Greenhouse Gas Regulation under the Obama Administration," *Harvard Journal of Law & Public Policy* 34, no. 2 (2011): 450.
29. Jonathan H. Adler, "A Conservative's Approach to Combating Climate Change," *The Atlantic*, May 30, 2012, 76.
30. Adler, "Legal and Administrative Risks."
31. Rothbard, *For a New Liberty*, 29.
32. Nozick, *Anarchy, State, and Utopia.*

33. Jeffrey Miron and Pedro Braga Soares. "What Should Policymakers Do about Climate Change?" Cato Institute Briefing Paper No. 130, *Cato Institute*, Nov. 30, 2021.
34. Ibid., 4.
35. See, e.g., F.A. Hayek, "The Use of Knowledge in Society," *American Economic Review* 35, no. 4 (1945).
36. Ibid., 5.
37. Dolan, "Controlling Acid Rain."
38. See, e.g., Roy E. Cordato, "Market-Based Environmentalism and the Free Market: They're Not the Same," *The Independent Review* 1, no. 3 (1997).
39. Robert W. McGee and Walter E. Block, "Pollution Trading Permits as a Form of Market Socialism and the Search for a Real Market Solution to Environmental Pollution," *Fordham Environmental Law Journal* 6, no. 1 (1994): 51–77.
40. Anderson, "George Bush Environmentalist."
41. Ibid. (emphasis added).
42. Ibid.
43. Richard A. Epstein, "The Principles of Environmental Protection: The Case of Superfund," *Cato Journal* 2, no. 1 (1982).
44. The Comprehensive Environmental Response, Compensation, and Liability Act, 42 U.S.C. §§ 9601 *et seq.*
45. Ibid., 10.
46. Ibid., 20–22.
47. See Ronald H. Coase "The Federal Communications Commission," *Journal of Law and Economics* 2 (1959): 29; Donald N. Dewees, "The Role of Tort Law in Controlling Environmental Pollution," *Canadian Public Policy/Analyse de Politiques* 18, no. 4 (1992): 429–431; Daniel H. Cole and Peter Z. Grossman, *Principles of Law and Economics*, 2nd ed. (New York: Wolters Kluwer, 2011), 397–400.
48. Epstein, "Principles of Environmental Protection: The Case of Superfund," 21–22.
49. Ibid., 33–34.
50. Ronald Hamowy, "Epstein's Theory of Environmental Protection," *Cato Journal* 2, no. 1 (1982): 39–44; Gerald P. O'Driscoll, Jr., "Pollution, Libertarianism, and the Law," *Cato Journal* 2, no. 1 (1982): 45–54.
51. Peter J. Hill, "Market-Based Environmentalism and the Free Market: Substitutes or Complements?" *The Independent Review* 1, no. 3 (1997): 395.

52. Ibid.
53. Daniel H. Cole, "Advantages of a Polycentric Approach to Climate Change," *Nature Climate Change* 5, no. 2 (2015): 116–117.
54. See Daniel H. Cole, "Political Institutions, Judicial Review, and Private Property: A Comparative Institutional Analysis," *Supreme Court Economic Review* 15, no. 1 (2007).
55. Adler, "Taking Property Rights Seriously: The Case of Climate Change," 213.
56. Ibid.
57. Ibid.
58. USTS 993, Art. 41.
59. See Daniel H. Cole, "The Problem of Shared Irresponsibility in International Climate Law," in *Distribution of Responsibilities in International Law*, eds. Andrei Nollkaemper and Dov Jacobs (Cambridge: Cambridge University Press, 2015), 307–311.
60. Ibid., 311 n. 86.
61. See "U.S. Climate Change Litigation: Common Law Claims," Sabin Center for Climate Change Law, (2022) http://climatecasechart.com/case-category/common-law-claims/.
62. See Albert C. Lin and Michael Burger, "State Public Nuisance Claims and Climate Change Adaptation," *Pace Environmental Law Review* 36, no. 1 (2018): 54–72.
63. Native Village of Kivalina v. ExxonMobil Corp, 696 F.3d 849 (9th Cir. 2012).
64. US Army Corps of Engineers, Relocation Planning Project Master Plan, Kivalina, Alaska, June 2006, https://www.poa.usace.army.mil/Portals/34/docs/civilworks/reports/KivalinaMasterPlanMainReportJune2006.pdf.
65. *Air Pollution and Control, U.S Code* 42 (2021) §§ 7401–7671q.
66. Massachusetts v. EPA, 549 U.S. 497 (2007).
67. *Urgenda Foundation v. The Netherlands* (2015) HAZA C/09/00456689 (June 24, 2015); aff'd (Oct. 9, 2018) (District Court of the Hague, and The Hague Court of Appeal).
68. Paris Agreement (Dec. 13, 2015), in UNFCCC, COP Report No. 21, Addenum, at 21, U.N. Doc. FCCC/CP/2015/10/Add, 1 (Jan. 29, 2016).
69. See, e.g., Graham Dawson, "Free Markets, Property Rights and Climate Change: How to Privatize Climate Policy," *Libertarian Papers* 3, no. 1 (2011): 19.

70. Epstein, "Principles of Environmental Protection."
71. 42 USC §§ 7401 *et seq.*, Part IVA.
72. See, e.g., Daniel H. Cole and Peter Z. Grossman, "When Is Command-and-Control Efficient? Institutions, Technology, and the Comparative Efficiency of Alternative Regulatory Regimes for Environmental Protection," *Wisconsin Law Review* (1999): 933–934.
73. See, generally, ibid.
74. Ibid., 908–909.
75. Daniel H. Cole, *Instituting Environmental Protection: From Red to Green in Poland* (Aldershot, UK: Macmillan and St. Martin's Press, 1998).
76. See, e.g., Brian Murray and Nicholas Rivers, "British Columbia's Revenue-Neutral Carbon Tax: A Review of the Latest 'Grand Experiment' in Environmental Policy," *Energy Policy* 86 (2015): 674–683.
77. U.N. Treaty Collection, Chapter XXVII Environment, 7.1, "Kyoto Protocol to the United Nations Framework Convention on Climate Change," December 11, 1997, https://treaties.un.org/Pages/ViewDetails.aspx?src=TREATY&mtdsg_no=XXVII-7-a&chapter=27&clang=_en.
78. Michael Grubb, Tim Laing, Thomas Counsell, and Catherine Willan, "Global Carbon Mechanisms: Lessons and Implications," *Climatic Change* 104, no. 3 (2011): 556–558.
79. Quirin Schiermeier, "Hot Air," *Nature* 491 (2012): 656–658.
80. David Victor and Danny Cullenward, "Making Carbon Markets Work," *Scientific American* 297, no. 6 (2007).
81. Cole, "Shared Irresponsibility in International Climate Law," 317. The EU's ban was effective because EU member states were the only major buyers of CDM credits once the US, Canada, and Australia had all pulled out of the Kyoto Protocol.
82. Cole, "Advantages of a Polycentric Approach to Climate Change Policy," 116.
83. Miron and Soares, "What Should Policymakers Do about Climate Change," 5.
84. Harold Demsetz, "Information and Efficiency: Another Viewpoint," *Journal of Law & Economics* 12, no. 1 (1969).
85. Albert Hirschman, *The Rhetoric of Reaction: Perversity, Futility, Jeopardy* (Cambridge, MA: Belknap Press, 1991).
86. Center for Biological Diversity v. National Highway Traffic Safety Administration, 538 F.3d 1172, 1200 (9th Cir. 2008).

87. See, e.g., Ronald H. Coase, "The Problem of Social Cost," *Journal of Law and Economics* 3 (Oct. 1960).
88. Martin L. Weitzman, "On Modeling and Interpreting the Economics of Catastrophic Climate Change," *The Review of Economics and Statistics* 91, no. 1 (2009).
89. Elinor Ostrom, *Understanding Institutional Diversity* (Princeton: Princeton University Press, 2005), 243.
90. Ibid., 256.
91. Daniel H. Cole and Elinor Ostrom, "The Variety of Property Systems and Rights in Natural Resources," in *Property in Land and Other Resources*, eds. Daniel H. Cole and Elinor Ostrom (Cambridge, MA: Lincoln Institute of Land Policy, 2012), 57.
92. Ibid.

Climate Change Adaptation Through the Prism of Individual Rights

David Dana

Introduction

Climate change has already imposed huge costs on the United States.[1] Climate-related costs can be contained to an extent by investments in adaptation, but who should bear the costs of those adaptation investments? This is a pressing question, because, by almost any calculation, some continuing climate change is now "baked into" the atmosphere, and mitigation efforts are lagging behind what many climate experts believe is needed to avoid climate change tipping points.[2]

This chapter explores the role of individual "rights" and the courts as individual rights vindicators in the coming fights over who should bear the costs of climate change adaptation.[3] In the liberal tradition, broadly understood, individuals and (sometimes) "legal persons" such as corporations have rights to use their property both for commerce and for their

D. Dana (✉)
Northwestern Pritzker School of Law, Chicago, IL, USA
e-mail: d-dana@law.northwestern.edu

J. H. Adler (ed.), *Climate Liberalism*, Palgrave Studies in Classical Liberalism,
https://doi.org/10.1007/978-3-031-21108-9_4

own non-commercial pursuits.[4] There are two strands of lawsuits that will test the question of how individual property rights mesh with the collective imperative of undertaking extensive climate change adaptation.

The first strand of suits involves property owners alleging that adaptation regulation deprives them of their property without just compensation, in violation of federal and/or state constitutional guarantees. We have seen, or (it is reasonable to posit) will see, suits by property owners regarding regulation addressing a range of climate-related problems, including sea-level rise, flooding from extreme weather, drought, and fire risks in populated areas.

The second strand of suits involves state and local governments seeking compensation from major greenhouse gas emitters like energy companies for the costs the states and localities will incur in adapting to climate change. These suits rely on a theory of "public nuisance," and in that sense could be thought to be outside an individual rights paradigm: public nuisance suits, definitionally, allege violations of a collective, rather than individual, right.[5] But the public nuisance label aside, these suits can be understood as about the limits of the individual rights of the defendant corporations. In essence, the defendants argue that they had individual rights—akin to property rights—to sell the products they sold (oil, gasoline, and electricity) because it was wholly lawful as a matter of positive law (both regulatory and tort law) for them to do so at the time of the sales. Hence, according to the companies and sympathetic pro-property rights NGOs, it would be wrong—a rights violation, even if the companies do not quite use rights rhetoric—for them now to be punished through the imposition of liability for the costs of adaptation.[6]

But individual rights are never absolute. Rights carry with them, and are circumscribed by, the responsibilities that attach to the rights. The responsibility not to cause harm through the otherwise privileged exercise of a right is generally recognized, even if what counts as "harm" is hotly contested.[7] In certain circumstances, our legal tradition constructively puts owners on notice of a harm-limitation on their property use when there is *no* actual, specific notice regarding the harm and the need for future regulation: any owner should understand that she has no right to engage in a property use if and when it comes to be understood that such use will cause drastic, immediate societal harm. To flesh out this intuition, consider the example offered by Justice Scalia in *Lucas v. South Carolina Coastal Council*[8]:

> It seems to us that the property owner necessarily expects the uses of his property to be restricted, from time to time, by various measures newly enacted by the State in legitimate exercise of its police powers .... [such as when] the corporate owner of a nuclear generating plant ... when it is directed to remove all improvements from its land upon discovery that the plant sits astride an earthquake fault.[9]

If climate adaptation regulation is aimed at property owners' behavior that will have effects akin to a nuclear explosion, then it seems irrelevant whether the owner had specific notice of harm that could result from her property use at the time they acquired title—of course, no compensation should be due. But in general, adaptation regulation is addressed at harm from property use that in and of itself, in the individual property owner's case, is not so extreme as to be socially intolerable under any circumstances. Thus, in the context of property rights suits against the government, a key issue will be whether the owner was aware or should have been aware of specific limitations in her title that would limit her exercise of her rights if doing so in fact became harmful in the previously-specified ways. If the owner was so aware or should have been, she will be hard pressed to argue that anything was really "taken" from her when regulation is implemented to address the harm.

As climate-related harms come to be better understood and more readily observable, property owners' argument that specific notice was lacking when they took title may become more difficult to maintain: No one building in Miami today can be said to lack notice that future, climate-related regulations affecting the use and value of their buildings may well be imposed in the next decade. Moreover, land use regulators and others may ensure (and in some places, already are ensuring that) that land purchasers and developers have specific notice of possible climate impacts and adaptation measures by including express conditions in building and other required permitting.[10] In that sense, even as the extent and economic burden of adaptation regulation increases (with the increase in climate impacts), property rights claims for compensation may become more legally tenuous.

At the same time, because governments almost certainly will have to pick and choose between which areas should be defended vigorously against climate impacts and which should not, the discourse may well take on the flavor of equal protection and anti-discrimination law, with property owners arguing along the lines of "why did my neighbors several miles away get to keep their beach house while I did not" or "why did the government protect that neighborhood from flooding but allow mine to flood?"

The inequality inquiry, in turn, will often implicate questions of causation. If owners can show that they would have been better off if the government had done nothing at all, they will be in a strong position to argue that the unequal treatment they received actually caused them property-damage harm. But if the rain (or wind or fire or sea surge) would have caused as much or more damage even if the government had done nothing, then the courts might construe the unequal treatment as simply the equivalent of non-actionable government inaction.

Questions of causation, albeit different ones, also will (or perhaps, more aptly, should) dominate the litigation brought by governments against major emitters like the energy companies. Some critics of the contemporary incarnation of public nuisance as a tort characterize it is a genuinely strict liability tort that can be used to impose liability even on the most careful and well-intentioned actors.[11] But the climate-related public nuisance suits allege that defendants (and the energy industry generally) knew their products were absolutely harmful, they deceived regulators and the public about those harms, and this deception accounts for—made possible—the social harms which the governments are now seeking to redress.[12] What remains very much to be seen is whether the plaintiffs can prove, or even will be afforded by courts an opportunity to prove, their allegations regarding actual deception and its causal consequences.

If plaintiff governments can succeed in proving that the defendant's deception is causally responsible for climate change and its effects, they will face another causation hurdle—proving that an ascertainable amount of damages are causally attributable to the particular defendants over whom the court has jurisdiction in the particular case, as opposed to those entities over whom the court does not have jurisdiction. And even as to the defendants before the court in the particular case, there will be thorny questions of how to allocate causal responsibility among them.[13]

In sum, both adaptation-related claims brought by property owners and those brought by governments implicate individual rights, although in quite different ways. In the end, the individual rights claims of property owners will depend on questions of questions of harm, notice, unequal treatment, and causation, and the individual rights defenses of the energy companies will (if the suits are allowed to proceed) depend on questions of deception and causation.

As this chapter discusses, an important commonality in both strands of litigation will be the centrality of questions of causation. Courts in both strands of litigation will be called upon to think through the appropriate standard(s) for causation in the climate adaptation context and to grapple with whether the fact-intensive adjudication causation disputes often necessitate is worth the legal system's resources. With respect to the latter question, I suggest that courts should be mindful of the large possible social benefits of such fact-intensive adjudication: whatever the actual litigation results, litigation discovery and even trials over causation could provide useful information about both what government and private actors have or have not done to either contribute to climate change and/or to manage a reasonable, equitable adaptation response to it.[14]

From the perspective of the proper vindication of individual property rights in the climate context, and the fundamental conception of fairness underlying our societal recognition of those very rights, I also suggest that courts in both strands of litigation consider "notice" and "causation" as related, rather than independent, inquiries. In the case of property owners suits against the government, the more pre-investment notice owners had that they might be subject to climate-related harms and the more notice they had that the government would or could not act as an absolute protector against any such harms, the more courts should insist on a high standard of causation with respect to the question whether the government "caused" a loss in the value of private property such that the government is liable for damages. Conversely, in the suits brought by governments against the major energy companies, the more pre-sale-of-products notice the companies had that their products could cause climate-related harm and the more the companies intentionally concealed that knowledge from the government and public, the less courts should demand on a high standard of causation that the companies caused climactic harms and thus should be liable for some of the costs governments bear as a result of the need to invest in climate adaptation. A flexible, variable standard of causation is appropriate in both contexts. As discussed below, however, the cases to date generally have not conjoined the notice and causation inquiries and have not openly embraced a variable standard for causation.

Property Owners' Takings Claims Based on Adaptation Regulation

The claims of property owners against governments will arise in an area of law—Takings law—that is at once highly developed and famously confusing.[15] This part very briefly reviews Takings law, and then illustrates how questions of harm and notice may be deployed by and debated among judges in addressing claims for compensation in adaptation-regulation contexts. The part then reflects on how property rights claims against the government ultimately may shift focus to questions of causation and discrimination, or at least unequal treatment.

Takings Doctrine in Brief

The federal and state Takings Clauses prohibit the taking of private property for public use without just compensation. When the government actually takes title to private property or physically takes over property without formally taking title, it almost always must pay just compensation. But when the government instead reduces the value of private property through regulation, the government generally need not pay any compensation. As Justice Holmes famously explained in the case that generally is thought to have created the category of "regulatory takings," government could hardly "go on" if it needed to provide compensation whenever regulations diminished property values.[16] The overwhelming number of regulatory diminutions in property value thus are non-compensable.

Property owners can prevail, however, in a few situations where the courts have held that a "per se" compensation requirement is applicable: namely, (1) permanent physical occupations of private property mandated by government regulation, as in a regulatory requirement that landlords must allow cable lines they do not own to be installed in their buildings[17]; and (2) 100% diminutions in value of a property due to the imposition of stringent regulations, as in a regulatory prohibition that prevents any activity or building on a bare parcel of land.[18]

Even these two per se compensation requirements have exceptions and are thus not, in truth, per se requirements at all. First of all, the government can reasonably condition the grant of a benefit—like the right to build in a zone where building rights expressly require regulatory permission—on the owner's granting some permanent physical occupation of the property by the government or the public where the occupation condition has a nexus to the harms the permitting or other regulatory restrictions are designed to prevent.[19] Second, the government can regulate away all the market value of a property interest if there is a "background principle" of law that should have put the owner on notice that such stringent restrictions might be imposed to prevent harm.[20] This heavy focus on notice comports with the essentially positivistic, rather than natural law, nature of our federal and state Takings Clauses: those clauses protect (or at least compensate) owners when states alter their law and legal regime of property, but the clauses do not guarantee any particular set of substantive property rights that endure for all time.[21]

The various opinions in the Texas Supreme Court decision in *Severance v. Patterson*[22] illustrate how the courts will struggle with the concepts of harm and notice in determining when adaptation-based regulation goes "too far" and constitutes a Taking. In 2005, Carol Severance purchased three properties on Galveston Island's West Beach. Five months after her purchase, Hurricane Rita struck and moved the line of vegetation landward (which under Texas law, marks the public/private divide), such that the entirety of Severance's house is now seaward of the vegetation line. The State claimed a portion of her property was located on a public beachfront easement, that portion interfered with the public's use of the dry beach, and the structure on that portion needed to be removed.[23] And with some obvious justification: Pre-2005 Texas case law strongly suggested that when the sea rolls in and pushes the vegetation line landward, the public has an easement over all the land seaward of the new vegetation line. Moreover, when Severance bought the land at issue, she received a disclosure mandated by Texas' Open Beaches Law that explained "that the property may become located on a public beach due to natural processes such as shoreline erosion, and if that happened, the State could sue, seeking to forcibly remove any structures that come to be located on the public beach."[24]

Despite all the constructive and actual notice Carol Severance received regarding the public claim on land seaward of the vegetation line, the majority opinion in *Severance* found that the intrusion on her property rights was so problematic that it required abundantly clear, unqualified, notice—notice "going back to 'time immemorial.'"

The dissenters disputed that the notice to Severance was in any way inadequate, while emphasizing the State needed latitude to protect the public interest in the beaches. In dissent, Justice Medina explained that "[b]y implied prescription, implied dedication, or customary and continuous use, overwhelming evidence exists that Texans have been using the beach for nearly 200 years." The dissenters stressed that pre-purchase Texas case law provided Severance constructive notice that the public easement rolled in when the sea rolled in, and that she had expressly purchased with knowledge of the claim by the public to the shore and the possibility that that claim would override private property rights.[25] The dissenters also took issue with the majority's failure to account for the fact that the hurricane activity had reshaped the landscape of the beach, affecting various property interests, including the public's interest.[26]

The back and forth of the sort we see between the majority and dissenters in *Severance* regarding harm and notice undoubtedly will continue in beachfront as well as other climate-sensitive settings, and different courts will reach different conclusions, even with similar facts. But the regulatory state may also make it more difficult for judges to ascribe inadequate notice to private property owners; regulators may do so by adopting new law, new regulations, and new permitting conditions. Notably, a post-*Severance* Texas statute, while opaque, seems to give regulators greater latitude to determine where the vegetation line is in the wake of storm activity, which may allow regulators to in effect preserve public beach that otherwise might have been rendered private under the *Severance* decision.[27] Similarly, in California, the Coastal Commission appears to be using permit conditions that in effect build rolling public easements into the title held by owners of private land on the seashore.[28]

Similar considerations of harm and notice are at play in the cases regarding restriction on private landowners' construction of sea walls to keep the rising ocean out, known as "hard armoring." Many experts believe that hard armoring is harmful to beach ecology and contributes to overall beach erosion, even if particular walls may protect nearby structures for a time.[29] Courts in states such as Oregon, California, and North Carolina have upheld restrictions on private owners' armoring against Takings challenges, heavily relying on notice ideas.[30] The courts emphasize that no landowner in a beachfront area has a reasonable expectation against land loss from erosion, since land lost to erosion, as a matter of longstanding common law, is public land.[31] To the extent landowners "lose" land to erosion, it is a loss to the forces of nature and not a Taking by the State.[32] Moreover, express armoring restrictions sometimes have been included in the development and building permits that were issued to the owners (or buyers from those owners) who later argue such restrictions constitute a Taking.[33]

Inequality, Causation, and Property Owners' Takings Claims

Climate change adaptation may force the question of the interplay of property rights protection and unequal treatment because governments may be forced to choose which neighborhoods or areas that will be protected in the face of climactic impacts and those that are not worth (from a net social welfare and/or raw political calculus) trying to protect.[34] Unequal treatment, for example, could take the form of allowing beachfront owners in some areas to protect themselves against rising seas, while restricting such activities elsewhere. Unequal treatment also could take the form of large government investment in beach replenishment, flood control, and fire protection in some "high value" areas alongside little or no investments elsewhere. In the fraught context of water rights and water allocations, regulators might be tempted to cut allocations for farmers who grow high water-consumption crops (like almonds) in order to satisfy urban populations' most basic needs.

A key doctrinal point that will be at the fore of such unequal treatment litigation is that the government is generally understood to *not* be liable under the Takings Clause for failing to take action that protects property values; conventional doctrine is that the government only can take property in a constitutional sense through affirmative action, not inaction.[35] Some commentators and property rights advocacy groups have argued for recognition of a category of inaction takings or passive takings, but that theory is in tension with the language and history of the federal and state Takings Clauses: the word "take," in both its original and current meaning, would seem to imply action, not inaction. Even more important is that the category of passive takings would have no coherent limitation. As a practical matter, if the courts did extend Takings liability to property losses that resulted from government inaction, the courts would find themselves in the unenviable position of hearing potentially tens or even hundreds of thousands of cases and of being required to second-guess how governments must decide to use their necessarily limited resources. Because inaction is (by definition) everywhere all the time and can always be claimed to have consequences on property values, there would be no meaningful limit to the number of possible inaction-based Takings claims. It is difficult to think courts would want to go down that road.

Nonetheless, when the government invests to protect one neighborhood or area (including public as well as other private lands) from flooding or fire or another climactic impact with the result that another neighborhood or area is even more burdened than it otherwise would have been, property owners in the burdened area can argue that the government action should be understood as a Taking of private property without just compensation. For example, if, in the face of extreme weather and flooding, the government chooses to direct flood waters to one side of the river rather than to the other, there is at least a plausible claim that the government caused a physical taking via flooding—but, and this is a key point, only if the damage exceeded what would have been the case had the government not intervened one way or the other.[36]

The ongoing litigation over flooding in Houston from Hurricane Harvey illustrates the kinds of questions courts will need to confront in unequal treatment litigation. In the face of massive storms that overflowed reservoirs, the US Corps of Engineers took some actions to try to limit the damage from flooding. Downstream landowners whose land was flooded have sued, alleging that the government in effect worsened the downstream flooding to protect the upstream owners by delaying the opening of the reservoir gates. But upstream owners also have sued, claiming the government's actions (including not opening the gates sooner) caused more flooding upstream than otherwise would have been the case. In the litigation brought by the downstream owners, the trial court concluded that the owners had no property right against protection against the Hurricane waters, even though it is clear that the waters flowed downstream after the government opened the gates of its overflowing reservoir.[37] But in the upstream case, the trial court seems to have accepted the upstream owners' arguments that they experienced more flooding—and in that sense, were deprived a property right—because the Corps did not do enough to ensure that upstream properties would not be flooded and hence caused the flooding of plaintiffs' houses.

The appellate decision to date that addresses the causal responsibility question most directly involves the Hurricane Katrina-related flooding of the Ninth Ward in New Orleans, which is one of the city's poorer wards and which had a majority African-American population.[38] In St. Bernard Parish v United States, property owners claimed that the Corps of Engineers took their property when the Ninth Ward flooded. The property owners argued that the Corps had built and then inadequately maintained a major channel in New Orleans that directed flood waters to the Ward that otherwise would have flowed to other parts of New Orleans. The federal court of claims agreed, holding the United States liable. Reversing, the Federal Circuit held that the evidence at trial did not establish that the Ninth Ward was made worse by the aggregate of the government's anti-flooding projects—which included not only the channel but also the levee system. The federal Circuit also opined that the evidence at trial did not establish that the flooding of the Ninth Ward would have been lesser had there been no channel constructed at all.

One issue raised by the St. Bernard Parish litigation is what exactly is the standard of causation that will apply when the government's climate adaptation efforts allegedly disadvantage some set of property owners as compared to others.[39] The Federal Circuit in St. Bernard Parish seemed to adopt a strict, but-for standard of causation, but other courts might be persuaded to adopt a less demanding, "substantial factor" standard. Likewise, courts can take more or less expansive views of the set of relevant government actions in framing the inquiry of whether the government caused more harm on net than otherwise would have occurred. The federal, state, or local government's actions can be conceived as only those immediately preceding the property loss in question or all the actions in the relevant area for decades or, of course, any number of possible time frames in between.

In deciding these difficult questions about what causation actually means and requires, courts may have a number of considerations that motivate them, including (1) a desire for consistency between the causation standard in climate cases and those in other Takings and tort cases, (2) concerns that government responses to pressing situations such as extreme weather and fire could be chilled or hamstrung by fear of Takings liability, and (3) concerns that governments will act unfairly toward property owners and perhaps especially relatively unprivileged owners in the absence of the threat of Takings liability. As discussed below, with respect to the third consideration, courts should consider how much and how timely notice private owners had of climate-related risks and the limits of government support in choosing a causation standard that could result in more or less liability on the part of the government.

Even when plaintiffs in such suits lose, as in the Hurricane Katrina litigation, there may be a social "win": when the courts allow discovery and/or trial, government accountability can be enhanced by the information that is made public. Much of what is known about what governments did well and did not do well to address flooding risks during Hurricanes Katrina and Harvey is known because of the extended litigation. Perhaps that kind of information itself will be an impetus toward better governance.

Governments Suing Energy Companies for Adaptation Costs in Public Nuisance

In the climate adaptation suits, state and local governments have sued major energy companies, arguing that their products, when burnt, produced greenhouse gasses that contributed to climate change, which in turn has and will require the plaintiffs to adapt to such phenomena as sea-level rise.[40] According to the plaintiffs, these companies have facilitated the burning of fossil fuels that account for something like 20% of aggregate emissions.[41] In so doing, the energy companies, it is alleged, created a public nuisance, which is doctrinally defined as an unreasonable interference with a public right.[42] More governments may file suit in the next few years.

The governments' adaptation costs suits are linked to the property owners' suits against governments alleging Takings in at least two ways. First, if the governments' suits do succeed, the governments will more readily be able to fund compensation for property owners. Second, and more conceptually, the governments' adaptation suits, like the property owner suits, implicate the questions of the limits of individual rights. The energy company defendants argue that it would be a violation of their rights in effect to hold them liable for selling wholly lawful products in conformity with all existing regulations and state products-liability-tort law (which is limited to individual consumer claims for individual injuries, and which, for that and other doctrinal reasons, is probably inapplicable to the sale of the fossil fuel products at issue). The companies characterize the public nuisance suits as illegitimate end-runs around state products liability law, which, they argue, is the only conceivable (but, again, legally inapplicable) harm-based limit on their rights to sell their products without subsequent liability.[43]

In the context of lead paint, a few decisions have allowed public nuisance suits to proceed against product manufacturers because they allegedly deceived regulators and the public as to the risks inherent in the products the defendants sold.[44] Building on these decisions, the plaintiffs in the adaptation costs suits allege that the energy companies withheld information about climate change and actively nurtured public doubt about the reality of climate change. The plaintiffs allege that, because of this deception, shifts toward renewable sources of energy that otherwise would have happened did not happen, climate change worsened, and the state and local governments now face costs of adaptation now and into the future that otherwise would have been avoided.[45]

The central question in these suits—if they are allowed to proceed—will be whether plaintiffs can prove that there would have been truly dramatic shifts in policy toward curbing emissions on a global scale had the companies *not* engaged in deception. A key issue will be, what will be the standard for the causation plaintiffs must prove? It will be difficult for plaintiffs to prove that, but for the companies' deception, we assuredly would now have a low- or no-carbon economy. After all, even today, after all that is known about climate change and its visible effects, we still do not see that many voters in the United States pushing for dramatic curbs on emissions.[46] But if the standard of causation is simply that the companies' deception was a substantial (or perhaps just contributing) factor to the climate change situation we now find ourselves facing, then the plaintiffs would have a better case for establishing causation.

The courts' decision as to what proof of causation really means may depend on a number of factors, including: (1) their concerns about not watering down causation so much for public nuisance law purposes that public nuisance suits outside the climate context might engulf the courts, (2) their view as to whether the blameworthy conduct of the companies (if the courts see it as such) justifies a lower causation standard or, instead, is (as an analytic matter) a wholly separate question; and (3) their sympathy with the plaintiffs' aim to jolt the political process to better take account of climate change needs or, to the contrary, their hostility to plaintiffs using courts and liability to influence political discourse.

Moreover, there will be other difficult causation issues. First, there is the difficult causal question of how much of the claimed adaptation damages are due to climate change, as opposed to multiple other factors. There are today more fires in California because of climate change, but climate change is not the sole explanation: a legally-facilitated underproduction of housing has forced people to seek housing closer to forested areas. Water shortages in the United States are due in part to climate change, but also to an arcane, wildly inefficient water allocation and usage legal regime.

Second, the courts will need to grapple with the question of how much causal responsibility for damages the defendant as a group or collective has. And that inquiry is problematic because no climate change suit has as defendants every major participant in the global energy industry and/or every major emitter. Some causally responsible actors—again, even assuming we limit such actors to major producers, refiners, and marketers of fossil fuels, which is a contestable limitation—thus will be absent in every suit.

Third, it is not clear how the relative causal responsibility of the defendants in any given suit should be assessed. Information regarding each major company's fossil fuel-related activities and sales and hence responsibility for emissions may be incomplete or subject to interpretation. Moreover, the level of each company's responsibility for deceiving the public arguably should matter. For example, if Shell's products generated more emissions than Exxon's (assuming available data can show that), but Exxon was more active in sowing doubt about climate change through deception, which company has a greater responsibility for adaptation-related damages?

The easiest way for courts to avoid these difficult causation issues is to dismiss the suits as legally insufficient, and courts indeed have employed a range of jurisdictional, abstention, and preemption doctrines to do just that in the litigation to date.[47] But as with the property owners' suits discussed above, there would be social benefits to courts allowing at least discovery on questions of causation in these suits against major energy companies. Through the suits, the public could learn how companies themselves documented their attempts to influence public policy and opinion and what they knew about climate risks. Disclosures during and because of litigation could inform and energize political debates about the climate problem.

The Notice/Causation Nexus

In both strands of litigation, it seems reasonable to ask that courts consider notice and causation as connected inquiries. In some instances, a property purchaser has reasonable notice both that (1) the property at issue is worth what it is only because the government has invested to protect it against such risks as flood and fire and (2) given climate change, that protection may not be assured in the future. When that is true, it seems only fair to private property rights values to demand that the property owners meet a high standard of causation with respect to flooding or fires that the plaintiffs allege the government caused by not doing enough to protect their properties. By contrast, where purchasers lacked such notice, a relaxed causation standard may be more consonant with respect for private property values. If that much is correct, then there is a good argument that the claims court in the upstream flooding case erred. The property owners in that litigation bought into developments right above a reservoir in a flood-prone and often flooded area and, in at least some cases, did so well after the basic facts of climate change and its connection to extreme weather were well-understood. That should have featured in the court's causation discussion, but did not.

Notice and causation also should be considered in concert in the public nuisance suits against energy companies. To the extent those companies had notice of the climate-related harms their products would cause and denied the public and government that notice through withholding and suppressing the information, there would seem to be less of a property-rights-free-enterprise-fairness argument against the deployment of a relaxed or at least contextualized test for deciding whether the companies caused climate-related damages. If that is correct, it would seem to follow a court should not address the causation issue in this litigation at all until after there is discovery regarding what companies knew and disclosed or did not disclose and whether they engaged in deception. Yet the Second Circuit in *City of New York City v. Chevron* dismissed one of these suits before there was any discovery and in part on a lack-of-causation rationale.[48] Both because the information would be a valuable addition to the public debate about climate change, and because it is highly pertinent to (what I suggest should be) a conjoined notice/causation inquiry, these suits should not be dismissed without some opportunity for the plaintiffs to engage in relevant discovery.

Conclusion

Climate change is the most pressing environmental problem facing the planet, and, even as efforts to mitigate climate change continue, governments will need to invest heavily in climate change adaptation. As the product of collective human action, climate adaptation seems like a collective problem that in no way implicates individual rights. But individual rights and their limits are implicated by both strands of adaptation litigation that this chapter discusses. Question of harm, notice, inequality, deception, and, especially, causation will play an important role in this litigation, and the interplay between notice and causation should (but may not) be considered. Regardless of whether plaintiffs ever recover in substantial measure in either strand of litigation, the litigation has the potential to inform and influence political debates regarding climate change.

Notes

1. See "New Report: Economic Cost of Climate Change Strains U.S. Economy," *Environmental Defense Fund*, August 12, 2020, https://www.edf.org/media/new-report-economic-cost-climate-change-strains-us-economy. See also Dana Nuccitelli, "Climate Change Could Cost U.S. Economy Billions," Yale Climate Connections, April 6, 2021, https://yaleclimateconnections.org/2019/04/climate-change-could-cost-u-s-economy-billions/.
2. Damian Carrington, "Climate Tipping Points Could Topple Like Dominoes, Warn Scientists," *The Guardian*, Guardian News and Media, June 3, 2021, https://www.theguardian.com/environment/2021/jun/03/climate-tipping-points-could-topple-like-dominoes-warn-scientists.

3. This chapter does not address the extent to which rights discourse affects climate litigation to compel emissions reductions, or climate litigation outside the United States. For a broader overview of climate litigation and the scholarship regarding it, see Jacqueline Peel and Hari M. Osofsky, "Climate Change Litigation," *Annual Review of Law and Social Science* 16, no. 1 (2020): 21–38, https://doi.org/10.1146/lawsocsci.807.
4. Adam Winkler, "What Rights Should Corporations Have?," *The Wall Street Journal*, March 1, 2018, https://www.wsj.com/articles/what-rights-should-corporations-have-1519919444.
5. For a brief description of the history and theory of public nuisance, see David A. Dana, "Public Nuisance Law When Politics Fails," *Ohio State Law Journal* 83, no. 1 (2022): 61–115. See also Thomas W. Merrill, "Is Public Nuisance a Tort?," *Journal of Tort Law* 4, no. 2 (January 2011), https://doi.org/10.2202/1932-9148.1113.
6. See, e.g., Chris Kieser, "Can a Legal, Non-Defective Product Be a Public Nuisance?," Pacific Legal Foundation, February 5, 2015, https://pacificlegal.org/can-legal-non-defective-product-public-nuisance/. This article argues that public nuisance suits violate principles of due process. There is also a third strand of litigation we have already seem to an extent, which involves *private* landowners seeking compensation against emitting corporations for damages resulting from what (allegedly) was a climate-change-induced event, such as a major Hurricane. See *Comer v. Murphy Oil USA*, 607 F.3d 1049 (5th Cir. 2010); *Comer v. Murphy Oil USA, Inc.*, 718 F.3d 460 (5th Cir. 2013); Ronald G. Peresich, "Climate Change Litigation," *The Brief* 45, no. 4 (2016): 28–33, 32. This article analyzes the *Comer* cases.
7. Mark D. Rosen, "When Are Constitutional Rights Non-Absolute? McCutcheon, Conflicts, and the Sufficiency Question," *William & Mary Law Review* 56, no. 4 (2015): 1535–1611.
8. *Lucas v. S.C. Coastal Council*, 505 U.S. 1003, 1027 (1992).
9. *Lucas*, 505 U.S. at 1027–1029.
10. As a leader in the climate change arena, California has been a focal point for integrating climate adaptation with permitting. See "Improving State Permitting for Local Climate Change Adaptation Projects," Little Hoover Commission, June 2017, https://lhc.ca.gov/report/improving-state-permitting-local-climate-change-adaptation-projects.
11. See, e.g., Donald G. Gifford, "Public Nuisance as a Mass Products Tort," *University of Cincinnati Law Review* 71 (2003): 741–837; Donald

G. Gifford, *Suing the Tobacco and Lead Pigment Industries Government Litigation as Public Health Prescription* (Ann Arbor, MI: University of Michigan Press, 2010); Victor E. Schwartz and Phil Goldberg, "The Law of Public Nuisance: Maintaining Rational Boundaries on a Rational Tort," *Washburn Law Journal* 45 (2006): 541–583; Victor E. Schwartz et al., "Game Over? Why Recent State Supreme Court Decisions Should End the Attempted Expansion of Public Nuisance Law," *Oklahoma Law Review* 62, no. 4 (2010): 629–665; Victor E. Schwartz and Christopher E. Appel, "Government Regulation and Private Litigation," *Boston University Public Interest Law Journal* 23 (2014): 185–218.

12. See Dana, "Public Nuisance Law When Politics Fails."
13. The courts have employed "market share" approaches at times to allocate responsibility where a certain kind of product with several manufacturers harmed people, but only with hesitation, and in the climate context, the number and identity of relevant market actors and their relevant shares is more contestable. Indeed, the relevant shares of emissions attributable to the defendants in the current suits brought by California and other governments already has been the subject of debate. On the other hand, newer methodologies may allow causal attribution with greater accuracy. Rupert F. Stuart-Smith et al., "Filing the Evidentiary Gap in Climate Litigation," *Nature Climate Change* 11 (2021): 651–655, https://www.nature.com/articles/s41558-021-01086-7?utm_source=other&utm_medium=other&utm_content=null&utm_campaign=JRCN_1_DD01_CN_NatureRJ_article_paid_XMOL.
14. In both contexts, the litigation and courts' decisions may impact the larger political debate—the debate in the legislatures—over climate change and responses to it. For example, if the courts were to suggest that property holders' claims had some validity, legislators and regulators might be more likely to either include some explicit or de facto compensation as part of adaptation regulation and/or they might work harder to configure the regulation such that its direct impacts on property values would be lesser. One possibility is that regulatory restrictions will be accompanied with transferrable development rights, which can temper the economic impact of regulations. Daniel J. DePasquele, "A Pragmatic Proposition: Regionally Planned Coastal TDRs in Light of Rising Seas," *The Urban Lawyer* 48, no. 1 (Winter 2016): 179–207. For a very thoughtful discussion of how nuisance and other tort litigation (through discovery and otherwise) can generate information to

inform and guide solutions to the climate crisis, see Professor Catherine M. Sharkey's excellent contribution in this area.

15. See, e.g., *Bridge Aina Le'a, LLC v. Hawaii Land Use Comm'n*, 141 S. Ct. 731, 731 (2021) (Thomas, J., dissenting from denial of certiorari) ("Our current regulatory takings jurisprudence leaves much to be desired."); *Nollan v. California Coastal Comm'n*, 483 U.S. 825, 866 (1987) (Stevens, J., dissenting) ("Even the wisest of lawyers would have to acknowledge great uncertainty about the scope of this Court's takings jurisprudence.").
16. *Pennsylvania Coal Co. v. Mahon*, 260 U.S. 393, 413 (1922) ("Government hardly could go on if, to some extent, values incident to property could not be diminished without paying for every such change in the general law. As long recognized, some values are enjoyed under an implied limitation, and must yield to the police power.").
17. See *Loretto v. Teleprompter Manhattan CATV Corp.*, 458 U.S. 419 (1982) (recognizing a per se compensation rule for regulations that effect a physical taking).
18. *Lucas*, 505 U.S. at 1003.
19. See *Nollan*, 483 U.S. at 825 (Upholding as constitutional permit conditions entail permanent access as long as the conditions have a nexus to the reasons for the regulatory restrictions); *Dolan v. City of Tigard*, 512 U.S. 374 (1994) (Affirming *Nollan* but also adding a reasonable proportionality requirement). One way to read the Supreme Court's highly problematic decision in *Cedar Point Nursery v. Hassid*, 141 S. Ct. 2063 (2021), is as implicitly constraining the universe of instances in which a government may require even minimal, temporary physical access to a business property without compensation as a condition of allowing that business to operate; the majority opinion suggests a compensation requirement applies even when temporary access serves the goal of ensuring the business will operate successfully and in the public interest.
20. *Lucas*, 505 U.S. at 1031.
21. David A. Dana, "Not Just a Procedural Case: The Substantive Implications of Knick for State Property Law and Federal Takings Doctrine," *Fordham Urban Law Journal* 47 (2020): 591–622. This article argues that federal Takings claims "rely directly and unavoidably" on uncompensated transitions in the positive law of a given State. But see Eric R. Claeys, "Takings, Regulations, and Natural Property Rights," *Cornell Law Review* 88 (2003): 1549–1671. This article argues that natural law has and should to a greater degree inform Takings jurisprudence.

22. *Severance v. Patterson*, 370 S.W.3d 705 (Tex. 2012).
23. *Severance*, 370 S.W.3d at 712.
24. *Severance*, 370 S.W.3d at 720.
25. *Severance*, 370 S.W.3d at 738.
26. *Severance*, 370 S.W.3d at 751.
27. For an explanation of how H.B. 3459 amends the Texas Open Beaches Act, see Megan McLaurin, "New Texas Open Beaches Act Amendment Explained," Surfrider Foundation, https://www.surfrider.org/coastal-blog/entry/new-texas-open-beaches-act-amendment-explained.
28. See, e.g., *Lindstrom v. California Coastal Commission*, 40 Cal. Ct. App. 5th 73, 252 Cal. Rptr. 3d 817 (2019). This case addresses a variety pf permitting conditions, including ones related to removal of structures and restrictions on armoring. See also Carolyn Ginno, "Do Mess with Texas? Why Rolling Easements May Provide a Solution to the Loss of Public Beaches Due to Climate-Change Induced Landward Coastal Migration," *San Diego Journal of Climate & Energy Law* 8 (2017): 225, which suggests California implement a permitting strategy. In addition to the problem of the inclusion of specific adaptation-related conditions in permitting, landowners suing for adaptation-related Takings also will have to contend with the fact that in many contexts, adaptation regulation both benefits and burdens the property at issue, such that the net just compensation due, if any, may be close to zero. See generally: Jacqueline Peel and Hari Osofsky's "Sue to Adapt?" *Minnesota Law Review* 99, no. 6 (2015): 2178–2250, for an explanation of how New Jersey and other case law in effect reduces the monetary value of Takings claims directed at adaptation restrictions by requiring that the benefits to the affected parcels also be included in the determination of just compensation.
29. See, e.g., Molly L. Melus and Margaret R. Caldwell, *Managing Coastal Armoring and Climate Change in the 21st Century* (Palo Alto, CA: Stanford Law School, 2015), CalCoastArmor-FULL-REPORT-6.17.15.pdf (stanford.edu); Jessica Grannis, *Adaptation Tool Kit: Sea-Level Rise And Coastal Land Use: How Governments Can Use Land-Use Practices To Adapt To Sea-Level Rise* (Washington, D.C.: Georgetown Climate Center, 2011), Adaptation_Tool_Kit_SLR.pdf (georgetownclimate.org).
30. See *Lindstrom*, 40 Cal. Ct. App. 5th at 73; *Nies v. Town of Emerald Isle*, 244 N.C. App. 81, 780 S.E.2d 187 (N.C. Ct. App. 2015); *Stevens v. City*

of Cannon Beach, 854 P.2d 449 (Or. Ct. App. 1993); *Shell Island Homeowners Ass'n v. Tomlinson*, 134 N.C. App. 217, 517 SE2d 406 (N.C. Ct. App. 1999).

31. *Nies*, 780 S.E.2d 187.
32. *Nies* at 200.
33. *Nies* at 200.
34. These choices already are the subject of mainstream journalism. See, e.g., Patricia Mazzei, "A 20-Foot Sea Wall? Miami Faces the Hard Choices of Climate Change," *N.Y. Times* (August 21, 2021), https://www.nytimes.com/2021/06/02/us/miami-fl-seawall-hurricanes.html; Christopher Flavelle, "With More Storms and Rising Seas, Which U.S. Cities Should Be Saved First?," *N.Y. Times* (June 19, 2019), https://www.nytimes.com/2019/06/19/climate/with-more-storms-and-rising-seas-which-us-cities-should-be-saved-first.html; Erica Goode, "A Wrenching Choice for Alaska Towns in the Path of Climate Change," *N.Y. Times* (November 29, 2016), https://www.nytimes.com/interactive/2016/11/29/science/alaska-global-warming.html.
35. See *Moden v. United States*, 404 F.3d 1335, 1345 ("The government's liability for a taking does not turn, as it would in tort, on its level of care. Instead, takings liability arises from an "authorized activity."); *All of Descendants of Tex. Land Grants v. United States*, 37 F.3d 1478, 1481 (Fed. Cir. 1994) ("A claimant under the Fifth Amendment must show that the United States, by some specific action, took a private property interest for public use without just compensation.").
36. In *Big Oak Farms, Inc. v. United States*, 131 Fed. Cl. 45, 48 (2017) the Corps of Engineers did exactly that, directing overflow waters during heavy rains to farmland on one side of the river to protect a small town (Cairo, Illinois) on the other side. The federal claims court held that the farmers whose land was flooded could prevail in a Takings claims but only if it could be established that the government's actions made the flooding worse than it would have been had there been no government flood control infrastructure and management at all.
37. *In re Downstream Addicks*, 147 Fed. Cl. 566, 583 (2020) ("In sum, there exists no cognizable property interest in perfect flood control against waters resulting from an Act of God," and "the Fifth Amendment does not make the Government an insurer" against flooding on a plaintiff's real property "when the government fails to completely protect against waters outside of its control").

38. See *St. Bernard Parish Government v. United States*, 887 F.3d 1354 (Fed. Cir. 2018).
39. Another unresolved issue is how the so-called emergency exception to Takings liability will be interpreted in the context of government efforts to address and contain the effects of climate change. A very old line of cases holds that, in a public emergency, a government can destroy some property without paying compensation if it does so to contain the emergency from destroying other property and endangering lives. There are plausible arguments that the old precedents drawn from sporadic urban fires and contaminated orchards and the like do not speak directly to a long-term problem like drought, fire and flooding associated with climate change. See Jeremy Patashnik, "The Trolley Problem of Climate Change," *Columbia Law Review* 119, no. 5 (2019): 1273, suggesting the emergency exception to Takings liability will, but should not apply to climate change adaptation scenarios. See also *TrinCo. Inv. Co. v. United States*, 722 F.3d 1375 (Fed. Cir. 2013) (extending the emergency exception to wildfires and specifically ones resulting from intentionally lit back-fires, but also suggesting more stringent requirements for the exception to apply). Moreover, actions like building sea walls in some areas but not others do not fit with the time-is-of-the-essence nature of the emergency exception.
40. In part because of the United States Supreme Court's decisions displacing federal common law regarding any sort of claim related to air pollution, these suits are proceeding under state law. See generally Jonathan H. Adler, "Displacement and Preemption of Climate Nuisance Claims," *Journal of Law, Economics and Policy* (forthcoming), retrieved at Case Legal Studies Research Paper No. 2021-14, https://ssrn.com/abstract=3950715. For a database listing the suits with links to the relevant materials, see "U.S. Climate Change Litigation, Common Law Claims," Climate Case Chart, Sabin Center for Climate Change Law, 2022, http://climatecasechart.com/case-category/common-law-claims/. The complaints in these lawsuits are substantially similar; the complaint filed by San Mateo County, for example, is almost identical to those filed by the other jurisdictions. See Complaint, *The County of San Mateo v. Chevron*, No. 171V03222 (S. Ct. July 17, 2017), [hereinafter San Mateo Complaint], http://blogs2.law.columbia.edu/climate-change-litigation/wp-content/uploads/sites/16/case-documents/2017/20170717_docket-17CIV03222_complaint.pdf. Sher Edling, LP is the law firm for the plaintiffs in all these suits to date.

41. See San Mateo Complaint, 33.
42. See American Law Institute, *Restatement of the Law (Second) Torts* (St. Paul, MN: American Law Institute Publishers, 1965), § 821B.
43. These views are well-expressed in a recent Chamber of Commerce's amicus brief urging affirmance of the dismissal of a suit by New York against major energy companies. See Brief for the U.S. Chamber of Commerce Litigation Center as Amicus Curiae Supporting Petitioners, *City of New York v. Chevron Corp.*, 993 F.3d 81 (2d Cir. 2021), https://www.chamberlitigation.com/cases/city-new-york-v-chevron-corp.
44. See, e.g., *People v. Conagra Grocery Products Co.*, 227 Cal. Rptr. 3d 499, 529 (2017) (relying heavily on *County of Santa Clara v. Atlantic Richfield Co.* 137 Cal. App. 4th 292, 302 [2006]).
45. See San Mateo Complaint, 62.
46. See Frederick Hewitt, "Why Wasn't Climate the Electoral Lever We Thought It Would Be?" WBUR, December 1, 2020, https://www.wbur.org/cognoscenti/2020/12/01/climate-change-joe-biden-frederick-hewett. Hewitt reports that only 4% of polled general election voters listed change as their top priority.
47. See Benjamin Ewing and Douglas A. Kysar, "Prods and Pleas: Limited Government in an Era of Unlimited Harm," *Yale Law Journal* 121, no. 2 (2011): 350. The authors claim that "in order for such dignified and pedigreed prompts [to the political system] to emerge, courts will have to avoid the temptation to run for political cover" in climate suits). See Melissa Mortazavi, "Tort as Democracy: Lessons from the Food Wars," *Arizona Law Review* 57, no. 4 (2015): 931. Mortazavi states: "Rather, in the overall context of the modern American legal landscape, tort law may be best understood as playing acritical balancing role in supporting democratic deliberation. Tort suits bring forth new ideas, create new forums for debate, force fact-finding, and increase back and forth dialogue amongst the public and private institutional actors to develop sound law and policy."
48. *City of New York*, 993 F.3d at 81.

Common Law Tort as a Transitional Regulatory Regime: A New Perspective on Climate Change Litigation

Catherine M. Sharkey

Introduction

Common law tort is often compared and contrasted with public statutes and regulations, as theorists map out and debate the factors that advantage one regulatory response to health and safety risks over the other.[1] Whereas tort law was the dominant regulatory regime in the nineteenth and early twentieth centuries, with the increase in legislative activity in the twentieth century[2] and the rise and proliferation of administrative

Danielle Altchiler (NYU Law 2023) and Michael Beckwith (NYU Law 2023) provided excellent research assistance. I benefitted from robust engagement at the Zoom conference with participants, especially Jonathan Adler, Karen Bradshaw, David Dana, Monika Ehrman, and Danny Shahar. For further comments, I thank Don Gifford, Greg Keating, and Rachel Rothschild.

C. M. Sharkey (✉)
New York University School of Law, New York, NY, USA
e-mail: catherine.sharkey@nyu.edu

J. H. Adler (ed.), *Climate Liberalism*, Palgrave Studies in Classical Liberalism,
https://doi.org/10.1007/978-3-031-21108-9_5

regulations in the latter half of the twentieth and into the twenty-first century addressing environmental, occupational, and product safety risks, tort law often recedes into the background.[3] To some, tort law is an anachronistic regulatory mechanism, inefficient and not well suited to regulating complex, cross-border national (and, increasingly, international) risks. Rather than respond head-on, stalwart defenders of common law tort typically claim it as the last bastion protecting against regulatory capture and legislative deadlock.[4]

What is surprisingly missing from most of the debate is a dynamic model of interactive regulatory responses evolving over time. Tort law can play an especially valuable role at the outset to fill a regulatory void, especially in areas that pose emerging and incompletely understood health and safety risks. Moreover, tort law can set in motion experimentation with various risk-minimization methods, with different jurisdictions embracing various measures to avoid, mitigate, and/or adapt to risks of harm. In that way, via a dynamic trial and error process, additional information regarding both risks and innovative techniques for mitigating and/or adapting to those risks is produced and can be assessed. Over time, with the additional information and learning from both successful and failed strategies, optimal regulatory approaches might emerge, which could then be enshrined in more uniform regulations.

Tort Law in the Administrative State: A Dynamic, Information-Forcing, Experimentation Model of Common Law Tort as a Transitional Regulatory Regime

The rise of the administrative state altered the course and trajectory of the common law of torts. While some have argued that tort is obsolete in light of the superior risk-reducing capabilities of regulation,[5] in practice, common law tort liability has persisted alongside federal regulation, including in highly regulated areas such as prescription medical devices and drugs, automobiles, workplace safety, as well as the environment.[6]

Scholars have at times strained, however, to provide a sound theoretical justification for common law tort liability, particularly in such highly regulated domains. Setting to one side the compensatory rationale for tort (which could readily be satisfied via a social insurance mechanism) and focusing exclusively on the regulatory function of tort law, the deck seems stacked against common law tort when compared with the expert regulator who also serves as a data aggregator.[7]

Common law tort has been touted as a "gap filler"—at the ready if and when regulatory enforcement fails. Under this conception, tort law takes a decidedly second seat to administrative regulation, but swoops in to confront harm-producing activities that have fallen through the cracks of an existing regulatory regime.[8] Professor Tom Merrill, however, disparages this longstanding role for tort law in the environmental realm, staking out an emphatic position that "[n]o serious claim can be made that [common law tort] public nuisance can serve a useful role as a fallback in the event of an emergency not covered by one or more administrative schemes."[9]

Perhaps in recognition of the flimsy grounding of the gap-filler role, scholars shifted to articulate the role of tort law as a necessary "catalyst" to stimulate new legislation and administrative regulation.[10] Professor Merrill grudgingly concedes: "This is the only plausible justification for the one remaining use of public nuisance in contemporary America—as an arrow in the quiver of claims asserted by state and local governments seeking to recover huge damage awards from corporate defendants allegedly responsible for various social ills inadequately addressed by existing regulation."[11]

What unites both the "gap filler" and "catalyst" conceptions of common law tort is a shared assumption (often implicit) that, all things equal, legislation and administrative regulation is preferable to common law tort, at least with regard to regulating complex modern health and safety risks. Where tort law is needed is to supplement when such administrative regimes fall short, or else (relatedly) to prompt further (stronger) regulatory responses. These conventional conceptions of common law tort are subject to flank attack—by pro-regulatory enthusiasts who see no need for the common law tort in a well-functioning modern administrative state as well as by anti-regulatory libertarians, who lament the diminished role for common law tort as mere gap filler or catalyst, subordinated to an ever-growing unaccountable administrative behemoth.[12]

Conspicuously absent from these conventional accounts (and critiques thereof), however, is the affirmative role for common law tort as a transitional regulatory regime, especially when facing indeterminate risks and great uncertainty with regard to optimal regulatory responses.[13] Tort (an ex post regulatory mechanism) is superior to regulation in this transitional stage, precisely because the health or safety risk is in the process of being fully understood, such that a regulatory agency (an ex ante regulatory institution) would struggle to craft an optimal response to the newly emergent risk. Common law tort can produce critical information—at the pleading, discovery, and remedial experimentation phases—that agencies can then use to hone their responses to such indeterminate risks.[14]

This information-generating role of common law tort is present from the commencement of a tort lawsuit. And, while it is not unique to the conception of tort in a transitional role,[15] it functions in a distinctive manner. With regard to global climate change litigation, public nuisance plaintiffs seeking damages and/or abatement note in their pleadings various mitigation and/or adaptation plans (past, present, and future), thereby making a public record of potential remedial strategies regulators can observe and collect data from.[16] Further insights gained during discovery would help regulators understand the scope of the emerging risk and its causal link to realized harms.[17] It would also shed light on plans being generated by state and local regulators for responding to risks that may not have been disclosed in the initial pleadings.

While filing and discovery can instigate information production, remedial action in favor of plaintiffs has even more to offer in terms of information generation. State court remedial action would allow regulators to see how enforcement experiments by common law judges can help craft the optimal regulatory response. Such remedial action is nonetheless not required for agencies to gain valuable insights into how responses have succeeded, and as importantly, how they have failed.[18]

There are benefits to letting common law tort take the regulatory lead in certain situations, in particular where the scope of emerging risks and its causal link to harms are poorly understood ex ante and where it is unclear what an optimal enforcement and/or remedial regime would look like. This is not to say such a system would be without costs. Surely there are some risks that do not lend themselves to state-by-state common law experimentation. Nonetheless, the benefits of such experimentation have received scant, if any, attention in existing theories of common law tort that appears to subordinate common law tort to agency regulation (absent regulatory failure of some kind).

Professor Merrill remains unduly skeptical: "Today, no one would think of responding to a new type of risk by urging the legislature to label the risk a public nuisance and entrust prosecutors and judges to use litigation to eliminate the risk."[19] But one need look no further than the canonical *Boomer* case for inspiration on this front. In that case, seven neighboring landowners claimed nuisance on account of property damage from dirt, smoke, and vibrations emanating from a large cement plant. At that time (1970), cement production had been "recently... identified as a significant source of particulate contamination in the Hudson Valley," as a type of air pollution "not only... damaging to the plaintiffs, but also... decidedly harmful to the general public."[20] In taking up the case, the court duly recognized that air pollution was a social problem that well exceeded the judicial ken.[21] Indeed, as the court recognized, the amelioration of air pollution was "likely to require massive public expenditure and to demand more than any local community can accomplish and to depend on regional and interstate controls."[22] The court realized that its courtroom was not the place "to lay down and implement an effective policy for the elimination of air pollution," which it conceded was "an area beyond the circumference of one private lawsuit."[23] Specifically, the court understood that "[e]ffective control of air pollution is a problem presently far from solution even with the full public and financial powers of government. In large measure adequate technical procedures are yet to be developed and some that appear possible may be economically impractical."[24]

But, notwithstanding the fact that it called for broader state and federal legislative and regulatory responses,[25] the court nonetheless proceeded to perform "its essential function" by adjudicating the private claims of the individual property owners before it.[26] Courts have often performed such a function prior to the emergence of statutory or administrative regulation in the environmental context.[27] Indeed, "environmental law was little more than tort law with an emphasis on nuisance" before Congress acted to vastly expand the role of the federal government in the 1970s.[28] Throughout the late nineteenth and early twentieth centuries, courts acted to regulate air and water pollution. Indeed, "[t]here is a rich history of common law cases that illustrate how people have protected themselves from pollution prior to passage of federal pollution legislation in the 1970s."[29] Courts took action, moreover, before scientific understanding enabled them to fully understand how and why pollution caused the problems it did to humans, livestock, crops, and land.[30] Tort law helps us discover what risks are valuable or important enough for us to care about and take action to address. Among other things, this helps explain why (at least in the environmental context) tort litigation of one sort or another often precedes regulation. When a risk becomes known and salient, someone tends to sue before a government starts to regulate.

In this way, common law tort plays a significant role in transitional periods. When *Boomer* was decided in 1970, the court acknowledged the role government was beginning to play in pollution control.[31] Federal regulation was just on the horizon, with the establishment of the federal Environmental Protection Agency (EPA).[32] Meanwhile, the court placed its faith in the information-forcing role of tort law: "It seems reasonable to think that the risk of being required to pay permanent damages to injured property owners by cement plant owners would itself be a reasonable effective spur to research for improved techniques to minimize nuisance."[33] Common law nuisance actions would generate information on a specific form of air pollution and use the threat of damages to spur ever improving research into pollution-reduction technology. Moreover, common law courts could experiment with remedial responses that might inform federal regulators down the road.[34]

Climate Change Litigation: Laying a Foundation for Experimentation with Adaptation Measures

Commentators are duly skeptical of common law tort as a regulatory response to risks of global warming. Indeed, regardless of where one stands on how great the current threat is, one would be hard pressed to disagree with Professor Richard Epstein that "if there is to be any attack on global warming it has to be done through a coordinated national program, not by piecemeal state actions."[35]

But perhaps here, too, we should pause to examine whether, notwithstanding the national (indeed international) scope of the risks, common law tort might play an information-forcing transitional role. Especially in light of great uncertainty with regard to the optimal strategies to mitigate and/or adapt to global warming, before rejecting out-of-hand the viability of "piecemeal state actions," it is worth exploring the extent to which such actions might encourage necessary experimentation in advance of the adoption of uniform federal regulation.

Moreover, adaptation strategies—as distinct from mitigation efforts aimed at limiting emissions of fossil fuels and other pollutants—inevitably address more localized effects of climate change, such as building sea walls or engaging in forest management strategies to address wildfires.[36] Thus, while climate change is a global phenomenon calling out for national (and international) regulation, its more localized impacts suggest a role for state or local action.[37] Moreover, "as we explore the best approaches to the problem of mitigation, we need to explore, with real urgency, how best to protect vulnerable people and places from what is now here, and very likely to get worse."[38]

The corpus of climate change litigation investigated here comprises sixteen causes of action brought by cities and municipalities, roughly evenly divided between East Coast[39] and West Coast[40] regimes (with one inland regime outlier[41]), against producers of fossil fuels. The lawsuits seek monetary relief to take proactive steps, including major infrastructure, to protect its residents and ward off the anticipated effects of global warming.

The conventional take on the utility of such lawsuits is that, by holding fossil fuel companies accountable, the lawsuits could succeed in reducing emissions directly (by requiring the defendants to cease or change emissions-producing conduct) or indirectly (through damage awards that bankrupt them or simply contribute to a weakening of their financial might so as to diminish their ability to lobby effectively against federal climate law).[42] Moreover, such lawsuits could also force the disclosure of information about the climate disinformation campaigns that is so gross that it tilts the public debate.[43] But, it would seem, at least in the mitigation context, the choice presented under state common law is somewhat binary, namely the companies are either liable or they are not. Thus, instead of experimentation and learning across jurisdictions, each jurisdiction has unilateral power because damages flowing from a liability holding in a single jurisdiction would likely prove massive in light of the scope of climate harms.[44] Here, instead, I consider whether and how the climate change litigation might impact the content of local adaptation policy, short of simply improving it by providing financial support in the form of a damages award.

The complaints and accompanying hazard plans in the climate change litigation provide a rich array of data and information regarding varied responses to the threat of global warming. Across the fifteen municipal and state actions,[45] mitigation and adaptation plans addressed key broad areas: electricity and energy; transportation/roads; water supplies and conservation; flood/storm prevention; structural protections to buildings; restoration of natural systems and natural resource protection; property protections and historic assets; public communication; public awareness/education; wildfire protection/prevention; public disaster/emergency services; community support; R&D; and waste prevention.[46] Experimentation will allow jurisdictions to build off of the successes and failures of other jurisdictions' courses of action when drafting their own plans; moreover, the existence of varied response strategies will lead to trial and error approaches toward developing optimal federal regulation.

To be sure, there is variation among the sixteen actions that reflects regional differences. Wildfire protection/prevention is a relevant category for West Coast areas, but absent from East Coast plans.[47] Certain plans leave out strategies for entire response categories, but these variations do not correlate with regional differences.[48] Other plans include unique strategies that may or may not reflect regional differences. Thus, the City of Boulder—which stands alone as an inland regime—hired a consultant to study alterations to ecosystems and species migration patterns due to a shifting climate in order to help the City understand public health risks.[49] The City and County of Honolulu included distinct strategies focused on affirming the island's cultural values and highlighting community support, prioritizing grassroots resilience, improving economic opportunity, and general island-wide alignment and cultural resiliency methods.

For my purposes here, the most significant variation is the state of development of the respective plan and, correspondingly, the amount of experiential information or empirical data supporting various mitigation and adaptation strategies. Plans range from very preliminary bare-bones mapping of potential strategies to fairly comprehensive, including information from already deployed successful (and failed) strategies.

At one end of the spectrum, are Anne Arundel and City of Annapolis, whose fledgling plans consist primarily of calls for more information so as to identify, develop, and create strategies.[50] Calls for more research and information bedevil plans in the earlier stages of development. Indeed, the majority of such plans are quite upfront about the need to conduct additional research in order to figure out the logistics of implementing more abstract goals and objectives. For instance, in the response category of electricity, the County of Maui lists "promoting and implementing policies like promoting energy and water efficiency and renewable energy" without any further description of how the County would go about implementing these policies.[51] In similar fashion, King County, Washington, details myriad plans to research and figure out the best methods rather than specific plans to implement specific strategies.[52]

At the other end of the spectrum, fairly comprehensive plans are set forth by Rhode Island, Baltimore, City and County of Honolulu, City of Oakland, County of San Mateo, and Delaware. With regard to Rhode Island and Baltimore, the majority of their mitigation strategies are fairly specific and actionable (especially as compared with the more vague mitigation principles outlined in the plans above). Honolulu similarly has a very developed plan, with a "Resilience Strategy" that outlines goals and strategies, accompanied by a "Multi-Hazard Mitigation Plan" that includes specific techniques to accomplish the wider goals. The plan, moreover, includes priority level and indicates that many of the strategies have already been implemented. Likewise, the City of Oakland has already deployed several projects to adapt to and mitigate the effects of climate change (e.g., Green Infrastructure Plan, Detain the Rain Program, Runway Safety Area Project, Repair the Middle Harbor Shoreline Park Dike).[53]

The first step toward learning is amassing information regarding potential strategies. Plans at later stages of development have already created strategies, worked out some of the kinks of implementing the strategies, and returned with a better understanding of which strategies are worth pursuing. Delaware's "All-Hazards Mitigation Plan" goes through each action item, the goals it supports, the category, the hazards addressed, lead agency responsible, estimated cost, potential funding mechanism, implementation schedule, and level of priority. Several other mitigation plans (e.g., County of Santa Cruz, County of San Mateo, and King County) likewise include lengthy appendices describing all of the mitigation and adaptation strategies the City or County has considered and whether the strategies have been rejected, completed, or incorporated into the existing plan.

The next step is evaluation—tracking the status and performance metrics for each strategy deployed. Delaware has tracked the status of all adaptation and mitigation actions around the state, compiling reports in 2007, 2010, 2013, and 2018.[54] County of Santa Cruz's "Local Hazard Mitigation Plan" has an appendix entitled "Successful programs and projects" which could be useful to see what strategies have been effective (and might be transferable to other jurisdictions).

A corollary here is to take particular study of failed efforts, namely strategies that were either abandoned or rejected. Delaware's plan (and several others noted above that catalogue strategies) lists failed efforts. In an appendix to its mitigation plan, the County of San Mateo lists each strategy that will not be carried over to the new plan, either because it is unactionable, is already included in an existing program, or is not feasible. An interesting feature is a column for strategies identified as "not feasible due to political and public backlash." This would be a particularly useful resource for other cities as they develop plans because it might be informative to understand which strategies were politically infeasible in another jurisdiction.

To date, careful review of the existing hazard mitigation plans reveals few examples of explicit reliance or borrowing from other plans. There were only two explicit cross references in the voluminous mitigation/adaptation plans associated with the sixteen lawsuits. The City of Annapolis referenced Anne Arundel County's Hazard Mitigation Plan. And while Annapolis did not explicitly state that it derived any of its own strategies from that plan, it cited it for the proposition that "changes to coastal and bay water quality caused by climate change may limit recreational and other water uses." A more cryptic reference to the "Hudson County Mitigation Plan" is included in Hoboken's mitigation plan.

But numerous municipal and state plaintiffs seem open to the potential of such borrowing—indeed several have requested funds as part of their litigation claims for research into other jurisdictions' strategies. Thus, the City of Charleston has requested funds to "investigate additional external best practices that may transfer to Charleston."[55] Charleston, meanwhile, has served as a model for the protection of property and historic assets. Annapolis has looked to Charleston (as well as New Orleans and Galveston) as part of its review of other historical communities to expedite building review and approval under disaster declarations.[56]

As part of its "Resilient Rhody" plan, Rhode Island aims to build relationships and learn from climate adaptation efforts in neighboring states to accelerate technical assistance to municipalities for local implementation. Its plan directs that "Rhode Island should look to the region for examples of how to model municipal and nonprofit partnerships that move from planning to prioritized project identification and funding." The state also looked further afield—to Florida—for guidance on hurricane-related strategies.[57] And the City of New York looked south to Louisiana and across the Atlantic to the Netherlands in search of potential models for amphibious homes—structures built on dry land atop buoyant foundations and pile supports, so they are able to float if a site is flooded—as it contemplates potential investments in these and other floating structures, designed to move vertically with tidal fluctuations and storm surge.

Jurisdictions could likewise compare the experience with regard to effective emergency communication strategies.[58] Charleston seeks to develop a web portal dedicated to informing citizens of flooding issues. Santa Cruz aspires to coordinate communication system upgrades with other agencies and cities, including evacuation operations for homes and businesses within specific areas.[59] The City of New York's well-developed system for public weather alerts could serve as a model. When dangerous weather conditions threaten, New York City monitors and disseminates the information it receives from the National Weather Service (NWS) and other sources.[60] Alerts like these provide New Yorkers time to take the necessary precautions needed in the face of impending coastal storms.[61] The City also sends weather advisory notifications to property owners, contractors, and developers to encourage them to take specific preventive actions when coastal storms threaten, such as removing or securing loose construction materials and items from sites that could become airborne in high winds.

In addition to borrowing strategies, jurisdictions might take a further step toward coordinating strategies. Many jurisdictions' plans include strategies involving local, state-wide, and federal coordination.[62] Annapolis, for example, is engaged with Rockefeller Foundation 100 Resilient Cities, and National League of Sustainable Cities parties to identify leading best practices for successful resiliency planning and implementation strategies.[63] And Baltimore's Disaster Preparedness and Planning Project includes the goal to "support and increase coordination and information sharing across jurisdictions to better enable mitigation of cross-border impacts on the region's watersheds (e.g., understanding flood conditions upstream in the county)," which would include partnering with local counties to evaluate major tributaries in all watersheds to determine best management practices for capturing run-off and slowly releasing it.

Conclusion: Common Law Tort as Part of the Future of Environmental Law?

As new environmental risks evolve—as they inevitably do—different legal tools are necessary to respond. Implicit in this chapter's conception of the dynamic transitional role for common law tort is a vision of state court judges using their common law powers to address emerging risks. A variety of remedial measures can be borne of these state court interventions, allowing agencies to see the different responses taken across the states and calibrate their regulatory response in turn.

With regard to global climate change litigation, we see—at least to a limited extent—that our federalist system encourages discovery through experimentation as different jurisdictions try different adaptation strategies, potentially learning from one another. What is less clear is the extent to which common law tort complaints against greenhouse gas polluters facilitate and/or enhance the incentives of states and municipalities to experiment with various mitigation and adaptation strategies.[64] Not only is the relevant baseline comparison difficult to discern but so is the extent to which the prospect of payments from fossil fuel companies accelerates or enhances this discovery. As the climate change litigation unfolds, moreover, there will be further discovery about legal strategies—e.g., which sorts of claims work—and perhaps even what settlements could look like, but, again, it is not entirely clear how that translates into tort as an engine as discovery for how to address and/or adapt to climate change harms.[65] One possibility is that fossil fuel companies might be spurred to innovate and experiment with what sorts of settlements they offer, agreeing to fund certain sorts of projects or others. Moreover, if settlements were structured around indemnification for certain sorts of harms, then the fossil fuel companies (or their insurers) would have incentives to help jurisdictions minimize actual climate risks and discover what sorts of adaptive interventions are most cost-effective. Hence, following the model seen in other areas of emergent risk, such as data breach, as a more robust third-party liability insurance market emerges in response to a greater threat of tort liability, insurers will engage in further risk management, exerting more potent regulatory control.[66]

Finally, a caveat. Scholars (and judges) who reflexively view newly emergent widespread risks as "too big" for state common law underappreciate the creative license traditional common law principles bestow upon judges.[67] Moreover, the increasing federalization of certain categories of tort lawsuits (including public nuisance) may exacerbate this trend, given the likelihood that federal judges are less likely to embrace the role of common law judge due to the restrictive domain of federal common law.[68] Central to the idea of tort as a transitional stopgap regulatory regime is that *state* high court judges exercising their common law powers will fashion such experimental remedies.

Notes

1. *See, e.g.,* Catherine M. Sharkey, "The Administrative State and the Common Law: Regulatory Substitutes or Complements?," *Emory Law Journal* 65, no. 6 (2016): 1705; Steven Shavell, "Liability for Harm Versus Regulation of Safety," *Journal of Legal Studies* 13, no. 2 (June 1984): 357.
2. *See generally* Guido Calabresi, *A Common Law for the Age of Statutes* (Cambridge, MA: Harvard, University Press, 1982), 5 ("[S]tarting with the Progressive Era but with increasing rapidity since the New Deal, we have become a nation governed by written laws.").
3. For example, the traditional narrative in environmental law suggests that tort was a precursor to the creation of federal environmental statutes, many of which essentially codified tort remedies. Environmental law casebooks tend to present private and public nuisance cases both for historical relevance—*i.e.*, from the "tort period" of environmental law—and to reintroduce tort principles that underlie statutory regimes such as CERCLA and OPA. This chapter challenges the evolution from common law tort to statutory regimes as a fixed, linear progression. I thank Karen Bradshaw for sharpening this insight.
4. Such defenses reach back into the early twentieth century when Roscoe Pound characterized the common law of torts as the last bastion of a democratic society. *See* Sharkey, "Administrative State," 1708 ("Pound inveighed against the dangerous accretion of centralized power and the expansion of a discretionary Executive Branch, which he claimed... threaten our balanced government and endanger the liberty of individual

Americans."). *See also infra* Note 14 (highlighting a comparative advantage of tort over legislative or regulatory processes in terms of airing and responding to non-majoritarian and/or well-funded positions).

5. *See, e.g.*, Thomas W. Merrill, "Public Nuisance as Risk Regulation," (unpublished draft), on file with author: 24 ("Public nuisance made sense at a time when the relevant risks were local and largely defined by custom, and government was a skeletal affair. With the emergence of an alternative mode of risk regulation in the form of the administrative state, the role of public nuisance as a type of risk regulation quickly became obsolete.").
6. *See generally* Catherine M. Sharkey, "Tort-Agency Partnerships in an Age of Preemption," *Theoretical Inquiries in Law* 15, no. 2 (2014): 359; *see also* Rachel Rothschild, "State Nuisance Law and the Climate Change Challenge to Federalism," *N.Y.U. Environmental Law Journal* 27, no. 3 (2019): 412.
7. Here I also set to one side a rights-based view of tort law—and only note here that, on different rights-based conceptions, common law tort may or may not have an especially significant role to play when novel forms of harm are at issue. *See infra* Note 26. In any event, by taking up tort law in its exclusively regulatory guise the deck is stacked most vociferously against tort—and thus presents the most formidable challenge.
8. *See, e.g.*, David A. Dana, "Public Nuisance Law When Politics Fails" (forthcoming), on file with author: 1–4 (observing that sometimes the "cracks" are more akin to chasms).
9. Merrill, "Public Nuisance," 21.
10. *See, e.g.*, Nora F. Engstrom and Robert L. Rabin, "Pursuing Public Health Through Litigation," *Stanford Law Review* 73, no. 2 (February 2021): 350.
11. Merrill, "Public Nuisance," 22.
12. *But see* Dan C. Shahar, "Libertarianism, Pollution, and the Limits of Judicial Adjudication", this volume (explaining that some problems are too large for the common law tort system to adequately remedy and libertarians should be more open to legislation and regulation in such situations).
13. In 2013, Thomas Merrill and David Schizer (implicitly) suggested that common law tort might play such a role with regard to the newly emergent risks associated with hydraulic fracking (e.g., potential contamination of groundwater), providing "a form of protection for

those injured by technological innovations, while information gradually accumulates that may eventually lead to more protective ex ante regulation." Thomas W. Merrill and David M. Schizer, "The Shale Oil and Gas Revolution, Hydraulic Fracturing, and Water Contamination: A Regulatory Strategy," *Minnesota Law Review* 98, no. 1 (November 2013): 145. *See* Catherine Sharkey, "Tort as Backstop to Regulation in the Face of Uncertainty," *JOTWELL*, November 26, 2013 (commenting that Merrill and Schizer's "proposals lay the groundwork for a more ambitious project: to reassess the balance between tort liability and regulation in areas that pose emerging, and incompletely understood, health and safety risks").

Fracking and consequent groundwater contamination provides a fruitful case study to examine the transitional role of common law tort. Namely, in 2022, what insights might be gleaned from a look-back over the past decade that considers the information-forcing role of tort litigation? I hope to pique the interest of scholars to take up this project.

14. Moreover, whereas majoritarian or well-funded positions are a prerequisite to federal legislation, minority and relatively poorly funded positions may still prevail in court. In this way, common law tort not only provides access to justice for minority positions lacking through legislative processes, but also provides a decentralized information-producing mechanism that arises from courts' necessary engagement with a broad array of positions (from the more radical to mainstream).

Ideally, the comparative advantage of tort vis-à-vis other institutional bodies (e.g., Congress, administrative agencies) should be subject to empirical scrutiny. *See, e.g.*, Sharkey, "Tort as Backstop" ("[Merrill and Schizer's] innate faith in tort law as a means to encourage risk-reducing innovation—based on their belief that 'products liability law has transformed the way consumer products are designed, and CERCLA has had a similar effect on waste disposal—wades into sharply contested waters that lack strong empirical foundations."). Moreover, when doing comparative institutional work, one must be careful not to compare an idealized version of tort law to a compromised, actual version of regulation and vice versa.

15. Both gap-filler and catalyst tort theories acknowledge the information-forcing attributes of common law tort. However, in such theories, this information-forcing role is a means to a different end. Gap-filler models see tort as either (a) seeking to produce information to highlight cracks in existing regulatory regimes addressing known risks or (b) acting as

a stopgap when political gridlock has left agencies without the capacity to adequately respond to known threats. *See* Dana, "Public Nuisance Law." Catalyst theories, while seeking to spur regulation and legislation, do not view the information generated by common law tort suits as a means to help agencies calibrate their responses for optimal deterrence. Rather, they seek mainly to highlight the existence of the problem to give political capital to those who want to regulate a given industry without regard for how that the regulatory response utilizes the information from the tort suit once the rule-making or legislative process has begun. *See* Engstrom and Rabin, "Pursuing Public Health."

16. Filing a lawsuit seeking damages for, or abatement of, an emerging threat might also raise public attention on the matter, leading in turn to the generation of information both inside and outside of the courtroom. *See, e.g.*, Engstrom and Rabin, "Pursuing Public Health," 355 (taking note of how opioid filings have generated revelations that were followed up on by the popular press).
17. More specifically, these cases might shed light on the role of industry in climate science and disinformation, which could be used by regulators, Congressional officials, and the public in structuring a response to the problem. For instance, could unearthing such relationships lead to changes in consumer protection statutes or new authority granted to EPA, similar to what happened in the wake of the cases against the tobacco companies? I thank Rachel Rothschild for raising this question.
18. Professor David Dana comes close to embracing this transitional role for common law tort. *See* David A. Dana, "Climate Change Adaptation Through the Prism of Individual Rights", this volume: 5 ("[W]hatever the actual litigation results, litigation discovery and even trials over causation could provide useful information about both what government and private actors have or have not done to either contribute to climate change and/or to manage a reasonable, equitable adaptation response to it."); ibid., 16–17 ("Disclosures during (and because of) litigation could inform and perhaps energize political debates about the climate problem, *what must be done about it*, and who should bear the costs.") (emphasis added).
19. Merrill, "Public Nuisance," 24.
20. Boomer v. Atlantic Cement Co., 257 N.E.2d 870, 876 (N.Y. 1970) (Jasen, J., dissenting).
21. Ibid., 871 ("The public concern with air pollution arising from many sources in industry and in transportation is currently accorded ever

wider recognition accompanied by a growing sense of responsibility in State and Federal Governments to control it."). Moreover, the dissent, which termed air pollution "a very serious problem in our State and Nation today," detailed how the New York State Legislature had enacted the Air Pollution Control Act "declaring that it is the State policy to require the use of all available and reasonable methods to prevent and control air pollution." Ibid., 875 (Jasen, J. dissenting).

22. Ibid., 871.
23. Ibid. The court elaborated: "Moreover, techniques to eliminate dust and other annoying by-products of cement making are unlikely to be developed by any research the defendant can undertake within any short period, but will depend on the total resources of the cement industry Nationwide and throughout the world. The problem is universal wherever cement is made." Ibid., 873.
24. Ibid., 871.
25. Ibid. ("It is a direct responsibility for government and should not thus be undertaken as an incident to solving a dispute between property owners and a single cement plant—one of many—in the Hudson River valley.").
26. Although throughout this book chapter, I maintain a regulatory perspective on tort law, a complementary story could be told from a rights-based lens; namely, *Boomer* recognized an emerging modern way of interfering with a long-recognized right to reasonable use and enjoyment of one's property. A new phenomenon, as it were, wedged into old legal boxes. Through this lens, public nuisance helps identify emerging forms of harm serious enough for tort law to regard them as rights invasions. I thank Greg Keating for pressing this point. That said, at least with respect to certain moral rights-based views, (too) much depends on one's definition of a "wrong"—and an exclusively historical focus could stymie the evolutionary growth of common law tort. *See* Catherine M. Sharkey, "Modern Tort Law: Preventing Harms, Not Recognizing Wrongs," *Harvard Law Review* 134, no. 4 (February 2021): 1423 (reviewing John C.P. Goldberg and Benjamin C. Zipursky, *Recognizing Wrongs* [Cambridge, MA: Belknap Press of Harvard University Press, 2020]).
27. *See* Richard J. Lazarus, "The Greening of America and the Graying of United States Environmental Law: Reflections on Environmental Law's First Three Decades in the United States," *Virginia Environmental Law Journal* 20, no. 1 (2001): 76 (describing how, prior to 1970, judges

employing "common law property and tort doctrines" were the only forms of legal environmental protection).

28. Alexandra B. Klass, "Common Law and Federalism in the Age of the Regulatory State," *Iowa Law Review* 92, no. 2 (February 2007): 566; *see also* City of New York v. Chevron Corp., 993 F.3d 81, 87 (2d Cir. 2021) ("The Clean Air Act was originally signed into law by President Lyndon Johnson in 1963 as 'the nation's first modern environmental law.' But it was not until seven years later, when Congress passed a round of significant amendments to the statute as part of the Clean Air Amendments of 1970, that the modern version of the Clean Air Act was born.").
29. Terry Anderson and Donald Leal, "Marketing Garbage," in *Free Market Environmentalism* (New York: Palgrave Macmillan, 2001), 138. *See, e.g.*, Susquehanna Fertilizer Co v. Malone, 20 A. 900 (Md. 1890) (enjoining a fertilizer company from emitting noxious vapors); McCleery v. Highland Boy Gold Mine Co., 140 F. 951 (D. Utah 1904) (enjoining a smelting company from polluting in a way that caused harm to crops and animals in the area); Whalen v. Union Bag & Paper Co., 101 N.E. 805 (N.Y. 1913) (enjoining a paper mill from polluting a river in a way that would have caused harm to neighboring farms) (cited in Klass, "Common Law and Federalism," 567–568).
30. Anderson and Leal, "Marketing Garbage," 138 ("[C]areful examination of common law pollution cases shows that judges were environmentalists long before regulators in national capitals.").
31. *Boomer*, 257 N.E.2d at 871 ("The public concern with air pollution arising from many sources in industry… is currently accorded ever wider recognition accompanied by a growing sense of responsibility in State and Federal Governments to control it.").
32. *E.g.*, William M. Blair, "Plan for Environmental Agency Gains," *N.Y. Times*, June 9, 1969, 34 (describing, inter alia, introduction of a "far-reaching bill to create a department of conservation and environment" in the Senate); "Case Urges New Agency To Protect Environment," *N.Y. Times*, June 1, 1969, 66 (commenting on Senator Case's introduction of a bill to create "a Cabinet-level Department of Environment to upgrade the fight against garbage, smog, and pollution."). *See City of New York*, 993 F.3d at 87 ("[T]he [EPA] was established in 1970 to implement programs to regulate pollution from both mobile and stationary sources under the Clean Air Act and other related statutes....").

33. *Boomer*, 257 N.E.2d at 873. The court, moreover, was not alone in seeing a role for tort law. *See* John W. Wade, "Environmental Protection, The Common Law of Nuisance and The Restatement of Torts," *Forum* 8, no. 2 (Winter 1972): 174 ("The fight for protection of our environment is a political one, involving the necessity of policy decisions at many levels. The courts can and do play a role, and private tort actions for damages or injunction can also play a significant part.").
34. In this regard, *Boomer* may not have fully realized its potential. *See, e.g.*, Daniel A. Farber, "The Story of Boomer: Pollution and the Common Law," *Ecology Law Quarterly* 32, no. 1 (2005): 127 ("[N]othing in the opinion indicates that… alternatives were even considered.").
35. Richard A. Epstein, "The Private Law Connections to Public Nuisance Law: Some Realism about Today's Intellectual Nominalism," (unpublished draft): 24.
36. *See* Karen Bradshaw, "Stakeholder Collaboration as an Alternative to Cost–Benefit Analysis," *Brigham Young University Law Review* 2019, no. 3 (2020): 715 ("Agencies are using a hybrid of analytic-data and collaborative analysis approach to *create localized climate response policies.*") (emphasis added). In addition to the scale of the issue (*e.g.*, local, state, federal, international), relevant factors may include the geophysical features of the pollution or resource being managed. For a provocative discussion of the latter, focused on social ecological systems theory effecting mechanism choice (albeit not considering common law tort), see Karen Bradshaw and Bryan Leonard, "Virtual Parceling," *International Journal of the Commons* 14, no. 1 (2020): 597.
37. *See, e.g.*, Br. for Appellants as Amicus Curiae, New York City Environmental Justice Alliance (NYC-EJA) and THE POINT, 3 ("[C]limate change, while experienced globally, is a problem with very local effects.").
38. Cass R. Sunstein, "Foreword: On the Imperative of Adapting to Climate Change," *Yale Journal on Regulation* 39 (2022): 478.
39. The eight East Coast regimes are: Anne Arundel County, City of Annapolis, City of Hoboken, Delaware, City of Charleston, Rhode Island, City of Baltimore, and City of New York.
40. The seven West Coast regimes are: City and County of Honolulu, County of Maui, King County, Pacific Coast Federation of Fishermen's Association (California), County of San Mateo, City of Oakland, and County of Santa Cruz.

41. The Board of County Commissioners of Boulder County (which includes City of Boulder and County of San Miguel) is the sole plaintiff from an inland regime.
42. *See* Thomas W. Merrill, "Is Public Nuisance a Tort?," *Journal of Tort Law* 4, no. 2 (May 2011): 48n192 (citing Amanda Little, "Public Nuisance No. 1: A Bold Lawsuit May Have Utilities Reconsidering Their Fight Against Regs," *Grist*, July 30, 2004, http://www.grist.org/news/muck/2004/07/30/griscom-lawsuit/) (arguing that the financial cost of litigation should make fossil fuel companies realize regulation is better for their profit margins that losses in court). The lawsuits might also encourage fossil fuel companies to support federal emissions control legislation that preempts state common law actions.
43. This is reminiscent of the type of information-forcing role suggested by "catalyst" tort theories. *See, e.g.*, Engstrom and Rabin, "Pursuing Public Health."
44. An email exchange with Katy Kuh helped clarify this distinction.
45. The Pacific Coast Federation of Fishermen's Association (a private plaintiff) did not include any reference to a mitigation plan and referenced minimal mitigation and adaptation strategies within the complaint. All of the mitigation and adaptation strategies for the PCFFA pertain to its crab industry.
46. In their complaints and hazard plans, the municipal and state plaintiffs categorized their mitigation and adaptation strategies in different ways. Most grouped the strategies either by hazard (e.g., flooding, earthquake, landslide), by goal (e.g., protecting human life, protecting private property, supporting emergency preparedness), or by type of technique (e.g., prevention, natural resource protection, emergency services).
47. For example, Maui's claim recites significant costs related to responding to increasingly severe wildfires, including two fires in July 2019 that burned 9200 acres and for which the Mayor declared a state of emergency, and a 4100-acre blaze that required forced evacuations and road closures in October 2019. The County provides firefighting personnel for such fires, which are becoming increasingly frequent and intense due to global warming.
48. For example, while Hoboken has not taken steps toward restoring natural systems and preserving natural resources, neighboring City of New York has developed many detailed strategies such as requiring that construction permits incorporate existing coastal-erosion management regulations to ensure that building activity will not accelerate

shoreline erosion. Similarly, while the Anne Arundel County mitigation plan includes several strategies targeting restoration of natural systems, the City of Annapolis does not include any strategies for this category.

49. "Warmer weather and shorter hibernation seasons could result in more human-animal exposure, including to species that typically carry rabies, such as skunks."
50. *See, e.g.*, Anne Arundel's "public disaster strategies for climate-related issues" category.
51. Additional examples are the strategies to "develop strategy to mitigate coastal risk" and "explore options and necessary code and regulation changes to allow greywater reuse systems for irrigation and toilet flushing." There are some more specific strategies as well (e.g., conduct a multivariate GIS analysis to identify and categorize open spaces for the County of Maui to purchase and preserve based on highest ROI for hydrological capacity and positive impact on flood mitigation) but the majority of these strategies still require additional research before implementation is possible.
52. For example, the complaint discusses that the County has incurred or will incur costs "associated with additional emergency planning, preparedness, response, and recovery actions" without any explanation of what the planning or strategies related to public disasters will entail.
53. These plans are highly developed—the City of Oakland has clearly researched and developed its strategies and is currently in the process of implementing them (though, like any project related to climate change, will need to continue to research the ways the strategies must evolve as time goes on).
54. For example, Delaware marked the flood-proofing of on-site residential septic systems located in the 100-year floodplain (2018 status: This action is currently being accomplished and supervised by the Division of Water Resources and Local government officials. There has been "limited success" as of 2018).
55. A particular aspiration is to collaborate with and learn Dutch and other methods of managing and living with water.
56. Annapolis has also looked closer to home, with stated plans to investigate and disseminate to businesses lessons learned in disaster recovery from Ellicott City, Maryland, and other Main Street Historic Districts.
57. Specifically, the plan's strategy to prioritize protection of the local fuel supply along evacuation routes during climate-related disasters (under the heading "Public Disaster Strategies for Climate-Related

Issues/Emergency Services") draws upon the experience during Hurricane Irma (2017) in Florida, when the large number of residents attempting to evacuate at the same time created a burden on the fuel supply, leaving many evacuees stranded without fuel, stalling their evacuation, and increasing calls for assistance from emergency services personnel.

58. Baltimore's plan calls for identifying "best practices" for the installation and management of floodproofing for all communications infrastructure at risk of water damage. And Rhode Island would expand into the realm of education, with a call to expand K-12 education on environmental literacy, including climate-related emergency preparedness, by developing resources for school use and identifying how these concepts can be incorporated into existing state standards.
59. Its mitigation plan includes an entry for an ongoing program to enhance support for interoperability communications system with local, state, and federal emergency services.
60. This includes NWS Watches and Warnings like a Storm Surge Watch, which indicates a possibility of life-threatening inundation from rising water moving inland from the shoreline, generally within 48 hours.
61. More specifically, New York City's "Know Your Zone Campaign" provides a variety of toolkits, planners, and digital aids so that residents can find their hurricane evacuation zone and understand how to prepare. And its "Ready New York" program has a branded emergency-preparedness initiative that includes a multilingual hurricane and coastal storm publication as part of its series.
62. Such efforts include working with surrounding jurisdictions, partnering with state governmental programs (e.g., in Charleston, coordinating with the South Carolina Department of Transportation), and working with federal research organizations.
63. It is also engaged with the U.S. International Committee on Monuments and Sites to identify leading practices internationally for public improvements, policies, and financing that supports sensitive flooding adaptation strategies for cultural sites.
64. Subnational governments certainly have incentives to develop adaptation plans—and even to compare notes on successful or failed strategies (e.g., Georgetown Climate Center's Adaptation Clearinghouse)—outside of

the litigation context. *See* "Legal and Policy Toolkits for Informed Decisionmaking," Georgetown Climate Center, accessed January 19, 2022, https://www.georgetownclimate.org/adaptation/toolkits.html.

In keeping with this chapter's theme of the interplay between tort and regulation, in order to tease out the information-forcing function of tort law, we should also consider the impact of state and federal environmental law, such as the National Environmental Policy Act and state variations, such as CEQA, on the development of adaption plans. Many of the municipal climate adaptation plans are not likely to implicate NEPA, or require environmental review even under state law, where such state environmental impact assessment law exists. In most instances, federal approvals are not required *for the plan*. If an individual adaptation action, such as elevating an interstate highway or some project involving Army Corps requires an approval and triggers NEPA, that review may have to look at other pieces of the adaptation plan, if they qualify as related or connected with potential cumulative impacts. Large scale projects like sea walls, or more likely, beach renourishment, may be the exception. And, in those instances, it may be that some pieces of an adaptation plan are able to benefit from the study put into a federal EIS. Thanks to Karen Bradshaw and Michael Burger for insights here.

65. Will local governments face the right incentives to carry out the most effective climate-related adaptation strategies? Or, instead, will they be driven to strategies designed to maximize payout? What if, absent the prospect of payments from fossil fuel companies, local governments would be more careful about only investing in those measures that will truly reduce risk? To begin to gain traction on this normative inquiry, the "failed" adaptation strategies uncovered in jurisdictions' plans could be further probed—first to discern how such efforts were deemed "failures" and whether it had any connection to legal strategy. I thank Jonathan Adler for raising this provocative point—and leave it to him (and others) to pursue.
66. *See* Catherine M. Sharkey, "Can Data Breach Claims Survive the Economic Loss Rule?," *DePaul Law Review* 66, no. 2 (Winter 2017): 384 ("There is emerging evidence that suggests many firms rate themselves as under-prepared for a data breach and that insurance companies are working with firms to provide some [albeit at this point limited] support aimed at preventing breaches [as opposed to acting primarily to mitigate losses in the aftermath of a breach]."); *see also* Catherine

M. Sharkey, "Revisiting the Noninsurable Costs of Accidents," *Maryland Law Review* 64, no. 1–2 (2005): 437–438 (describing the risk management function of insurance).

67. *See, e.g.*, City of New York v. Chevron Corp., 993 F.3d 81 (2d Cir 2021) (characterizing climate change as beyond the ambit of state nuisance law); Shahar, "Libertarianism" (advocating for legislation and regulation for issues such as climate change that are beyond the remedial capabilities of common law tort). Shahar may well be correct that common law—if conceived not as a transitional regulatory mechanism but instead as a regulatory substitute—will lead to underenforcement due to systematic constraints on the scope and reach of its enforcement function. *But see* City and County of Honolulu v. Sunoco LP, Civ. No. 1CCV-20-0000380 (denying oil and gas company defendants' motion to dismiss on preemption grounds by noting: "This is an unprecedented case for any court, let alone a state court trial judge. But it is still a tort case. It is based exclusively on state law causes of action.").
68. Samuel Issacharoff and Catherine Sharkey, "Backdoor Federalization," *UCLA Law Review* 53, no. 6 (August 2006): 1353; Catherine Sharkey, "A Quiet Revolution in the Work of the Federal Courts: The New Federal Common Law of Torts" (working paper).

Libertarianism, Pollution, and the Limits of Court Adjudication

Dan C. Shahar

Libertarians often exhibit hostility toward environmental legislation and regulation. Among the strictest libertarians, this manifests in blanket rejection of such interventions.[1] More moderate libertarians often concede legislation and regulation are sometimes needed in the face of environmental challenges, but only as non-ideal "second best" options.[2]

Libertarians who deprecate legislation and regulation do not intend to imply environmental problems should go unaddressed. Instead, they favor other means of resolving these problems. Libertarian principles leave broad latitude for undertaking private projects, entering voluntary governance arrangements, and bargaining with others to secure desired outcomes. Crucially, when environmental problems come from actions that violate rights, libertarians believe courts should defend victims and hold aggressors accountable.

D. C. Shahar (✉)
Tulane University, New Orleans, LA, USA
e-mail: dcshahar@gmail.com

J. H. Adler (ed.), *Climate Liberalism*, Palgrave Studies in Classical Liberalism,
https://doi.org/10.1007/978-3-031-21108-9_6

Court adjudication plays a distinctive role in libertarian environmental theorizing because of its power to *compel* people to alter their behavior. Although libertarians support private projects, voluntary governance arrangements, and bargaining efforts, they do not believe people are obligated to undertake such activities. Thus, libertarians cannot guarantee these pathways will prevent environmental destruction if people prefer not to participate. However, libertarians insist everyone is required to respect others' rights whether they wish to or not. Thus, courts provide a singular mechanism within libertarian theory for preventing strong-willed despoilers from running roughshod over the environment.

The judiciary's role in libertarian environmental theorizing is especially salient in the context of pollution. Pollution problems typically involve people harming those around them, so empowering the victims to defend themselves in court seems like a natural response. Some libertarians concede resolving complex, large-scale pollution issues like climate change, acid rain, and ozone depletion may be beyond courts' capabilities. But many view more mundane instances of pollution as soluble simply by charging judges with enforcing rights.

The standard libertarian analysis captures the vital truth that courts can often restrain pollution without the need for legislation or regulation. However, this chapter argues it has been a mistake for libertarians to regard common law adjudication as a general-purpose substitute for other forms of state intervention. Courts left to operate on their own will predictably provide inadequate protection for pollution victims. The most effective way for courts to fulfill the role libertarians envision is to work alongside legislators and regulators as complements instead of replacements.

The reason legislation and regulation are essential for controlling pollution has to do with the separation of powers in liberal societies. According to the traditional liberal framework, a judge's job is not to transform longstanding legal principles to keep pace with changing values and circumstances. Instead, it is to apply sound legal reasoning to cases brought before the court—including when the judge feels the law ought to be different from what it is.[3] Reforming the law amidst changing values and circumstances is the task of legislators and administrators. Libertarianism stands the best chance of addressing the challenges pollution creates by allowing each side of this arrangement to perform its role, not by substituting one for the other.

Before elaborating on these claims, I should clarify this analysis does not imply a blanket endorsement for environmental legislation or regulation. Many take it for granted such interventions are needed to protect citizens and promote social wellbeing. Yet, libertarians rightly caution against concluding directly from the shortcomings of private action and court adjudication that alternative interventions are needed. In practice, legislative and regulatory processes have drawbacks that cast doubt on their potential to deliver better outcomes than private actors and courts.[4] Sometimes, lawmakers and administrators even *undermine* environmental protections by siding with despoilers against their victims.[5] This chapter will argue courts' institutional limitations should make libertarians more open to environmental legislation and regulation. However, the wisdom of embracing these alternatives in practice depends on the merits of specific policies as well as our faith in political actors to make things better rather than worse.[6]

My discussion proceeds as follows. "Enforcing Rights" reviews standard libertarian narratives regarding the courts' role in securing environmental justice. "Navigating Indeterminacy" clarifies how such accounts can render nuanced, non-absolutist judgments about pollution despite significant theoretical indeterminacy. "Limited Protections" explains why, in the absence of legislation and regulation, courts that operate in the manner described in "Enforcing Rights" and "Navigating Indeterminacy" will deliver inadequate protection against many common pollution threats. "Institutional Competencies" links this inadequacy to the judiciary's traditional place in the liberal separation of powers and explains why legislation and regulation are essential for overcoming it. "Recasting the Judiciary" considers whether reconceiving the judiciary's functions could preempt my critique. "Focusing on Rights" examines whether libertarianism would retain a distinctive place in environmental debates if it embraced an expanded role for legislation and regulation.

Enforcing Rights

According to one model of libertarian political morality, property rights delimit a protected domain within which individuals ought to be secure against unwanted incursions. In Robert Nozick's famous formulation of this view, others' rights are not simply considerations to be weighed when deciding how to pursue our goals; instead, they represent "side constraints" on the range of options we may legitimately consider.[7] On its face, this understanding of libertarianism seems to imply a very strict anti-pollution outlook. As Morris and Linda Tannehill write:

> If I own the air space around my home, you obviously don't have the right to pour pollutants into that air space any more than you have a right to throw garbage onto my lawn. Similarly, you have no right to dump sewage into my river unless we have a contract specifying that you may rent the use of my river for such purposes (and that contract would have to include the consent of all those individuals who owned sections of the river downstream from me, too).[8]

However, characterizing rights as "side constraints" threatens to rule out far too much.[9] In modern societies, it is possible to trace virtually everything we do to at least some environmental impacts. Prohibiting *all* pollution as a matter of principle would effectively pit libertarianism against industrial civilization. It would hardly alleviate the issue to stress that polluters can validate their activities by gaining consent from all affected parties. As a practical matter, the costs of negotiating such agreements for every polluting activity would be prohibitive. And at a more fundamental level, it simply seems false we have a right to veto any action that produces even a tiny environmental impact on us. For better or worse, some degree of intrusion upon our living spaces is inevitable in social life. Libertarianism must accommodate this fact if we are not to reject it out of hand.

To their credit, many libertarians recognize the need to avoid such absolutist commitments. For example, Nozick concedes it is sometimes permissible to infringe upon rights provided we fully compensate those affected.[10] Others insist we must interpret individual rights to leave "elbow room" for others to live their lives,[11] with tiny trespasses allowed if they do no substantial harm.[12] Both caveats echo more mainstream legal traditions that distinguish between inviolable "property rules" and compensation-permitting "liability rules"[13] and hold that *de minimis non curat lex* ("the law does not concern itself with trifles").

Clarifications like these help us see how libertarians can square a general commitment to individual rights with non-zero levels of pollution. Contrary to the blunt assertion we have "no right" to pollute others' spaces without their consent, a more sophisticated libertarian analysis accepts such rights exist to at least some extent. According to this revised outlook, our duty is not to avoid pollution altogether but rather to keep it within tolerable bounds and compensate victims for any substantial harm done.

Navigating Indeterminacy

Non-absolutist interpretations of rights enable libertarians to oppose harmful pollution without ruling out every polluting act. Yet, these positions face challenging interpretive questions that do not arise for simpler models of inviolable "side constraints." For example, under what conditions do pollution impacts rise to a level where courts should intervene? When should polluters be allowed to compensate victims and continue polluting rather than being forced to stop?

Abstract ethical theory cannot tell us precisely how to resolve such puzzles.[14] However, history shows legal regimes can evolve answers through the accumulation of precedent. When individuals bring polluters to court for perceived wrongs, the task confronting judges is not to articulate fully general principles for handling pollution. Instead, it is to resolve the specific dispute before them. Each decision provides a template for adjudicating similar cases in the future. As precedents build upon each other over time, the results are nuanced legal standards for settling a dizzying range of controversies.[15]

As John Hasnas observes,[16] the emergence of these robust systems of practice is no mere coincidence: on the contrary, it would be surprising if it did not occur. Because courts offer a unique venue for pressing grievances, judges will repeatedly find themselves in a position to resolve pollution conflicts. Even if a specific judge fails to adjudicate a controversy satisfactorily, the likelihood similar disputes will recur means other judges will have opportunities to find better answers. Appealing to evolutionary legal processes thus gives libertarians a powerful way to address pollution concerns without relying on legislation or regulation.

It also bears repeating that highlighting courts' capacity to control pollution does not commit libertarians to the claim we must use this specific mechanism to resolve every pollution problem. As noted earlier, libertarians recognize broad rights to pursue private projects, form voluntary governance arrangements, and bargain with others. These pathways enable people to pursue very different environmental outcomes than would arise through legal adjudication. The reason courts figure so prominently in libertarian theorizing is thus not that they represent the sole pathway for addressing pollution. Instead, it is that courts provide victims with a potent tool for *compelling* polluters to stop, compensate, or both, depending on the details of the case.

Limited Protections

The discussion thus far has focused on the positive case for enlisting courts to control pollution. This section now explores a critical limitation of that argument. First, I argue legal evolution is susceptible to path dependency, and we can expect pressures operating in early industrial societies to produce legal practices that deny precautionary protection against pollution. Second, I show that the inherently uncertain nature of many pollution-mediated harms means victims who cannot appeal to precautionary standards will systematically struggle to defend themselves in court. Together, these considerations imply legal regimes where courts operate in the absence of legislation and regulation will tend to deliver protection with worrying gaps—even if judges perform admirably in adjudicating the cases brought before them.

Path Dependency

Let us begin with the phenomenon of path dependency. I have said abstract ethical theory provides non-absolutist libertarians with incomplete guidance on precisely when and how to limit pollution. Instead of deriving standards from the armchair, these libertarians must lean on evolutionary legal processes to generate strategies for addressing the diverse conflicts pollution creates.

There are few theoretical constraints on the types of legal standards these processes could generate.[17] We can imagine societies developing stringent standards that protect individuals from all but the most demonstrably benign pollution. Likewise, we can imagine looser standards where courts decline to act unless polluters impose appalling impacts. We cannot rule out any of these possibilities at the level of theory. But in practice, we should expect processes of legal evolution to be shaped by the social circumstances in which they occur. When it comes to pollution law, these processes will tend to unfold in a specific type of milieu: the emergence of industrialization.

There is a sense in which pollution problems are inseparable from communal living. People living alongside others inevitably confront smells, noises, and other annoyances for which legal adjudication provides a potential solution. However, there is another, narrower sense in which pollution is a novel problem that arises as societies begin experiencing economic development. Although there are respects in which smoke-filled skies and toxic rivers are analogous to unkempt pigsties and raucous gatherings, they have a very different societal meaning. The significance of the social upheavals that accompany industrial pollution is unlikely to be lost on court officials.

History confirms judges rarely step before the freight train of industrialization to insist polluters halt their activities.[18] Instead, precedent typically develops in ways that allow industries to operate without abandoning the idea of controlling pollution more generally. The experiences of Great Britain and the United States illustrate several ways this can happen.

One such approach effectively recasts nuisance law as a doctrine of reasonable behavior. Do polluters' activities generate sufficient public benefits to justify their burdens? Have the polluters made conscientious efforts to limit their impacts? If so, then pollution may be allowed to proceed, with victims meriting only compensation for specific, demonstrable injuries.

We can see this approach in a 1904 Tennessee case, *Madison v. Ducktown Sulphur, Copper & Iron Co.*[19] In this case, several residents of a rural mountain area sued a pair of mining companies that had been processing ore nearby. The miners' operation involved burning large "roast piles" of wood and ore to expel impurities. The landowners complained the smoke from these roast piles was harming their health, damaging their timber and crops, and generally ruining their lives. They thus asked the court to halt the mines' operations with an injunction.

Although the Tennessee Supreme Court sympathized with the residents, it was struck by the fact these two mining companies were effectively the *raison d'être* for the entire community in which the defendants lived. Before the mines' arrival, the area had supported only about two hundred people, and that number had subsequently grown to twelve thousand. The mining companies paid roughly half of the taxes collected in the county, compensated their employees to the tune of a half-million dollars per year, and consumed numerous goods and services provided by the local community. Meanwhile, the combined value of the plaintiffs' properties was approximately $700. Compounding matters, the court found the mining companies were conducting their operations in an entirely typical way—indeed, "the only known method by which these plants can be operated, and their business successfully carried on"[20]—and had "made every effort to get rid of the smoke and noxious vapors, one of the defendants having spent $200,000 in experiments to this end, but without result."[21]

To the court, there was no doubt the residents deserved compensation for the mines' demonstrable harms. However, they regarded the decision of whether to issue an injunction as less straightforward:

> The question now to be considered is, what is the proper exercise of discretion, under the facts appearing in the present case? Shall the complainants be granted, in the way of damages, the full measure of relief to which their injuries entitle them, or shall we go further, and grant

> their request to blot out two great mining and manufacturing enterprises, destroy half of the taxable values of a county, and drive more than 10,000 people from their homes? We think there can be no doubt as to what the true answer to this question should be.[22]

Thus, the court ruled the mines could continue operating provided they paid compensation for their impacts.

The *Madison* court employed a balancing test to assess whether pollution should be allowed to proceed, weighing its burdens against the social benefits of polluting activities. However, this was not the only form of reasoning courts used to reconcile themselves to environmental impacts in the era of industrialization. A different approach focused on the idea mentioned earlier that some pollution is inevitable in modern civilizations, and courts must not interpret citizens' rights so stringently as to rule out the possibility of industrial development.

This outlook is evident in an 1865 English case, *St. Helen's Smelting Co. v. Tipping*.[23] Mr. Tipping had come to own a large estate in an industrial area outside Liverpool, and a copper smelting operation subsequently came online just over a mile away. The smelting process emitted noxious gases that destroyed Tipping's hedges and trees, sickened his cattle, and generally made his estate a miserable place to live. So, Tipping brought the smelters to court to ask for relief.

Presented with this conflict, the English House of Lords distinguished the observable damage the smelters' fumes had done to Tipping's property from the personal discomfort Tipping had endured. On the second issue, the court noted Tipping had purchased an estate in an industrial area, and he could hardly expect the court to insulate him from the inconveniences of life in such a place:

> [I]n the county of Lancaster, above all counties, where great works have been created and carried on—works which are the means for developing the national wealth—you must not stand on extreme rights, and allow a person to say, 'I will bring action against you for this, and that, and so on.' Business could not go on if that were so.[24]

It was only the demonstrable physical damage to Tipping's property that grounded a valid legal complaint:

> [T]he submission which is required from persons living in society to that amount of discomfort which may be necessary for the legitimate and free exercise of the trade of their neighbors, would not apply to circumstances, the immediate result of which is sensible injury to the value of the property.[25]

The court thus awarded Tipping compensation for the damage to his vegetation and livestock, but it denied him relief for his discomfort and permitted the smelters to continue operating.

Two things about this decision are particularly noteworthy. The first is the court's contention that one must interpret the stringency of a person's rights with an eye to their community's character. As Lord Westbury commented on the case:

> With regard to ... the personal inconvenience and interference with one's enjoyment, one's quiet, one's personal freedom—anything that discomposes or injuriously affects the senses or the nerves—whether that may or may not be denominated a nuisance, must undoubtedly depend greatly on the circumstance where the thing complained of actually occurs. If a man lives in a town, of necessity he should submit himself to the consequences of those obligations of trade which may be carried on in his immediate locality, which are actually necessary for trade and commerce, also for the enjoyment of property, and for the benefit of the inhabitants of the town and of the public at large.[26]

The second is the sheer magnitude of inconvenience the court believed citizens should be prepared to accept in the name of commerce. Tipping's complaint was not simply that the smelters were emitting an unpleasant odor or some unsightly smoke. All parties to the suit agreed industrial facilities in Tipping's area had desolated the landscape—with the smelting company even appealing to this as a fact in their favor:

> The defendants' case is, that the district in which the plaintiff lives is almost without a living tree, and the hedges are universally blackened and destroyed. In point of fact, there is no doubt that the effect of the manufactories throughout the district is entirely destructive of vegetation.[27]

The court saw nothing inherently illegal about creating a situation like this in the name of development. In places where such pollution had become the norm, citizens were the ones who needed to adapt. It was only when polluters caused demonstrable physical property damage that individuals could properly appeal for relief in the form of compensation.

I do not intend to suggest via these examples that courts *never* stood up aggressively for victims against major industrial polluters, for that is not so.[28] Nor has my purpose been to exhaust the range of possibilities for how courts can reconcile industrial development with a commitment to protecting individual rights. Instead, my point is that in societies where rapid industrialization is lifting whole populations out of poverty, and where blocking pollution raises concerns about jeopardizing progress,[29] we should not be surprised when courts demand proof of serious harm before intervening. Correspondingly, we should not expect courts to require polluters to meet the types of precautionary environmental standards now common throughout the developed world, whereby even a substantial *risk* of harm is considered sufficient for intervention. Although there would be nothing incoherent at the level of theory about courts applying the latter types of standards, the practical circumstances in which pollution problems arise create powerful pressures in the opposite direction.

Uncertainty

I have argued that given the social context in which we can expect pollution law to develop, we should not expect courts to converge on strongly precautionary standards in the absence of legislation and regulation. With this claim in hand, we can turn to the second stage of my argument regarding the limits of court adjudication for controlling pollution. Here, I argue that when courts do not adopt precautionary standards for protecting victims from risk and instead focus on demonstrable harms, victims will systematically struggle to defend themselves in court against many common pollution threats.[30]

To see the issue, consider a case in which a complainant arrives in court ascribing a recent cancer diagnosis to a factory emitting noxious fumes near their house. In one version of this scenario, the complainant's specific form of cancer is virtually unknown except in individuals exposed to the factory's chemicals. Here, we might expect the court to readily find in favor of the complainant and, given the gravity of the diagnosis, to strongly consider enjoining the factory's operations unless it brings its emissions under control.

However, in a second and more typical version of the scenario, the connection between the complainant's diagnosis and the factory's pollution is not so cut-and-dried. Suppose the factory operators explain that, like most cancers, this illness could have many causes, including the complainant's diet, weight, alcohol or tobacco use, lack of exercise, history of viral infection, and genetics. They correctly note that standard estimates ascribe only a tiny fraction of cancers to environmental pollution.[31] The complainant struggles to demonstrate the factory's causal responsibility for her cancer in the face of these claims. Indeed, they cannot even provide community-level statistics that definitively show cancers like theirs occur locally at higher rates than could be expected from random variation. The best they can do is provide air quality test results showing the factory emits known carcinogens at substantial concentrations alongside nationwide scientific studies linking these chemicals to elevated cancer rates.

How a court decides this second kind of case will depend on the legal criteria for proving a rights violation. I have said that, in theory, such standards could be highly precautionary, holding that exposing a person even to considerable *risk* of harm violates their protected domain. A benchmark like this might push a court to intervene against the factory even without clear proof their activities had caused this specific illness. On the other hand, applying a legal standard that demanded convincing evidence of concrete harm would likely result in the dismissal of our complainant's accusations as unsubstantiated.

Similar dynamics exist for many other burdens that can arise through pollution exposure, e.g., asthma, coughing, nausea, headaches, kidney disease, heart disease, and stroke. Each of these ailments has many causes. In many situations, it will be statistically unlikely a specific firm's pollution is to blame for a given person's condition. Uncertainties compound further in areas where numerous firms contribute to local pollution in diverse ways.[32] If courts adopt precautionary standards, citizens in these contexts may be able to prevail despite their inability to establish clear causal links between their troubles and polluters' activities. But if courts impose more demanding epistemic standards, the predictable result will be that victims systematically struggle to defend themselves.

Predictably Limited Protection

When taken together, the tendencies discussed in this section imply sharp limits on the protection citizens can expect from courts pressed to operate on their own. Because legal standards concerning pollution are likely to evolve during periods of industrialization, we should expect courts to settle on standards that allow significant levels of pollution and recommend intervention only in response to clear demonstrations of substantial harm. Because of the inherently uncertain nature of many pollution harms, such standards will create hurdles for establishing victimhood that will prove insuperable in many common pollution contexts. There is nothing about libertarians' abstract theory of environmental justice that requires this result, and much of their rhetoric recommends against it. But even so, we should not be surprised when asking courts to control pollution without legislation or regulation leaves victims with significant gaps in protection.

In the industrializing societies in which these dynamics are likely to emerge, it may be deemed unproblematic for victims to struggle to block polluters: indeed, this may even be desired. But over time, industrializing societies tend to become wealthier and progressively less tolerant of sacrificing health and quality of life for even more material prosperity. As environmentally conscious citizens struggle to persuade courts to protect them from pollution, we may expect them to increasingly regard the choice to entrust pollution control to the judiciary as suspect. Yet, judges who honor their duties to uphold the law will hesitate to discard long-established legal practices even when they would personally prefer for the law to be different from what it is. The pressure to honor precedent contained in the dictum, *stare decisis* ("stand by what has been decided"), will present a significant obstacle to those who would charge courts with controlling the many difficult-to-trace harms pollution creates.

None of these concerns imply a regime based solely on court adjudication *cannot work*. Indeed, the development of pollution law in most industrial common law nations followed roughly the trajectory I have described until the emergence of modern environmental legislation and regulation in the latter part of the twentieth century. These earlier regimes were broadly functional. The problem was simply that their results became increasingly unattractive over time. Even as citizens became more concerned about the impacts of pollution, courts continued upholding standards of adjudication that emerged during earlier stages of industrialization. Self-described victims often found themselves powerless to block their perceived abusers. It is hardly surprising legislation and regulation were eventually demanded to redress the situation, yielding the robustly interventionist regimes we now see worldwide.

Anticipating demand for legislation and regulation does not imply an endorsement for specific interventions, whether historical or theoretical. As I noted at the start of this chapter, the wisdom of embracing concrete environmental policies depends on their merits, and there are often many reasons to doubt legislators and administrators will deliver genuine improvements. To yield a practical case for intervention, we would need to augment our discussion of the demand for solutions by assessing their supply.[33] However, it is not my goal to provide such an argument in this chapter. I seek only to highlight that when libertarians fixate on court adjudication to the exclusion of legislation and regulation, they neglect even the *possibility* these alternative means may be appropriate. Given the systematic drawbacks of the common law approach, this neglect is inappropriate, even if it will sometimes turn out to yield correct practical guidance.

Institutional Competencies

The preceding discussion cast doubt on courts' ability to pivot from permissive liability standards suitable for periods of industrialization to stricter standards more in line with the values of an affluent society. But why should we expect legislators and regulators to be more capable of undertaking this type of reorientation than courts? Why would we not expect courts to adapt their practices on their own?

There is no theoretical reason courts *cannot* manage the shift I have described. However, this section will argue potent institutional pressures work against this. Because of the judiciary's role in the traditional liberal separation of powers, judges who honor their profession's norms will tend to regard major intentional transformations of the law as beyond their institutional competency—even when they privately believe reforms would be desirable.

To begin to see the issue, consider the role traditionally ascribed to judges in liberal societies. Although judges plainly exercise legal authority, liberals have typically not viewed them as *originators* of the law's demands. Instead, judges are meant to *discover, interpret, and apply* the law rather than create it. For Montesquieu, removing the authority to create and implement law from the judiciary's domain was essential to prevent judges from becoming tyrants:

> [T]here is no liberty, if the judiciary power be not separated from the legislative and executive. Were it joined with the legislative, the life and liberty of the subject would be exposed to arbitrary control; for the judge would be then the legislator. Were it joined to the executive power, the judge might behave with violence and oppression.[34]

Many liberal nations' constitutions manifest Montesquieu's preferred division of state powers among legislative, executive, and judiciary branches. The justice professions in these nations have likewise internalized the idea that a judge's role is distinct from that of a legislator or regulator.

It is important not to exaggerate the extent to which judges can do their job without producing new laws. As we have seen, the mechanism by which common law standards arise in the first place is the accumulation of precedent. It is a point of logic that judges who initially establish precedents cannot do so by rote application of preexisting rules: some creativity is necessarily involved. We must also note judges sometimes depart from long-standing practices and overturn time-honored precedents they deem misguided. These facts notwithstanding, it remains true that when judges confront disputes, they do so under considerable institutional pressure to reconcile their judgments to established legal practice. This idea is captured in the previously mentioned maxim, *stare decisis*.

In the context of pollution issues, this pressure to conform to prevailing practices is bolstered by the fact that even when judges would like to hold polluters to stricter standards, it will typically be difficult for them to maintain earlier cases were decided *incorrectly*. Abstract theory provides no uniquely appropriate standard for resolving these cases, and I have already indicated precautionary practices are inappropriate for industrializing societies. Given these facts, a judge's conviction that new circumstances call for tightening standards displays the character of just that: someone's opinion. Not only would enforcing such an opinion in the face of longstanding legal norms fly in the face of the traditional role accorded to judges, but it would also generate reasonable complaints from polluters who may have gone to great lengths to comply with the norms. When we add that victims will generally have inconclusive evidence of the harms they allege—while large financial investments and numerous jobs may plausibly be regarded as hanging in the balance—it is not difficult to see why judges may be reluctant to substitute precautionary interpretations of what counts as a rights violation for existing standards that focus more narrowly on clearly demonstrable harms.

These considerations should lead us to expect judges to respond sluggishly to evolving public values that increasingly emphasize health and quality of life over continued economic development. Because of the role traditionally ascribed to courts in the liberal separation of powers, established precedents will tend to impose a substantial drag on processes of judicial innovation that might tighten pollution standards over time. The claim is not that courts *cannot* evolve their practices, for they undoubtedly can. But insofar as changing social conditions make it desirable to reform legal practices deliberately to strengthen protections against pollution, this task most naturally falls to the branches of government whose express functions are to create and implement laws—namely, legislatures and administrations.

By emphasizing this role for legislation and regulation, I do not mean to advocate *displacing* courts from the task of controlling pollution. Even with these other tools on the table, courts would retain essential roles in protecting individuals. For one thing, although my emphasis in the preceding sections has been on courts' limitations, there is no denying their effectiveness in many domains. Especially given the imperfections of real-world legislatures and administrations, we should expect courts to provide the best option for resolving many of the conflicts created by pollution—particularly those involving clear and demonstrable injuries or nuisances falling outside the norm for the areas where they occur.

Further, as Catherine Sharkey discusses in this volume,[35] one significant court function is to illuminate areas where legislation and regulation are needed. When victims seek justice in court and do not find it, this signals that other forms of intervention may be desirable. In this role, courts' fact-gathering processes generate invaluable information to guide the development of potential new legislation and regulation.

It is also possible to craft laws and regulations to operate in partnership with judges rather than push them out of the way. For example, instead of enacting environmental performance standards and enforcing them unilaterally, lawmakers and regulators can focus on specifying new systems of rights for courts to uphold. By supplying the kinds of precautionary legal standards courts are unlikely to evolve independently, governments can go a long way toward filling the gaps I describe in this chapter without assuming more comprehensive control over economic life.[36] We can say similar things about other administrative functions like supplying scientific expertise and data for private parties to use in adjudicating their conflicts.

Recasting the Judiciary

I have argued libertarians stand the best chance of adequately protecting those affected by pollution by taking advantage of all components of liberalism's institutional division of labor. Indeed, the template provided by liberalism's separation of powers makes this position appear nearly tautological. The entire idea behind the separation of powers is to make each branch of government reliant on the others. Within this framework, leaning on the judiciary while excluding the legislative and executive branches would inevitably leave essential functions unfulfilled. Insofar as libertarians accept a need to transform legal standards governing pollution as societies develop, and insofar as they grant it is not the judiciary's role to contradict long-standing precedents to enact desirable reforms, *something* like legislation or regulation seems essential.

For their part, however, libertarians may retort I have simply failed to understand the relationship between their political theory and the traditional liberal framework. Libertarians who want courts to control pollution without relying on legislation and regulation plainly cannot construe the judiciary's role in the way liberals have articulated it. Perhaps some of the functions liberals have traditionally ascribed to legislators and regulators should reside with the judiciary instead.

One way to accomplish this would be to explicitly authorize judges to reform and recalibrate legal norms to align with changing circumstances and values. However, although this would help libertarians dodge the worry I present in this chapter, it would do so only by raising serious concerns about the concentration of arbitrary power in judges' hands. Jurists unencumbered by expectations to conform to past decisions would be free to impose whatever arrangement they saw fit in each case coming before them. A society organized this way could not correctly be described as operating under the "rule of law." The intricate and stable system of legal practices I described as the great product of legal evolution could not exist in such conditions. Citizens would have little basis for planning their lives around the enforcement of any specific standards of conduct.

To avoid this undesirable conclusion, libertarians might note that imbuing *the judiciary* with additional powers does not require investing *individual judges* with those powers. For example, libertarians could charge the judiciary in a more holistic sense with crafting and revising legal standards to meet the needs of a changing society. A judicial body of this kind appears to be envisioned by Rothbard when he writes:

> [I]t would not be a very difficult task for Libertarian lawyers and jurists to arrive at a rational and objective code of libertarian legal principles and procedures based on the axiom of defense of person and property, and consequently of no coercion to be used against anyone who is not

> a proven and convicted invader of such person and property. This code would then be followed and applied to specific cases by privately competitive and free-market courts and judges, all of whom would be pledged to abide by the code, and who would be employed on the market proportionately as the quality of their service satisfies the customers of their product.[37]

For discussion's sake, let us set aside whether Rothbard is being realistic about the ease with which legal codes can be designed. The more important feature of Rothbard's vision for our purposes is its allusion to a body of legal authorities who provide courts with the standards they enforce. Insofar as such a body could revise and clarify the law amidst changing conditions, judges would not need to concern themselves with these tasks at the bench. This delegation of authority would preempt concerns about judges taking expanded mandates as an invitation to wield arbitrary and abusive power.

By broadening the judiciary's powers, libertarians could dodge my arguments about the limits of its institutional competency. However, they would do so only by reintroducing the roles of *legislator* and *regulator* in a different form. Rothbard's body of "Libertarian lawyers and jurists" would manifest a very different pathway to legislation and regulation than are familiar in liberal societies. But insofar as this body possessed the power to reform legal practices and specify appropriate pollution standards, it would be engaging in legislation and regulation in the sense relevant here.

This section's discussion highlights that legislation and regulation do not *necessarily* need to be carried out by distinct parliamentary bodies and administrative agencies. Insofar as libertarians hope to escape the specific pathologies of such entities, it is open to them to advocate alternative institutional pathways for fulfilling similar functions. However, the point stands that something important is missing from libertarian theories that treat court adjudication as a *substitute* for legislation and regulation. Especially in changing circumstances that render long-standing legal practices substantively objectionable, societies have strong reasons to be open to legislation and regulation *in some form* to provide up-to-date standards for judges to enforce.

Focusing on Rights

Suppose libertarians concede that legislation and regulation play essential roles in controlling pollution and hence are not merely "second best" alternatives for when other options fail. At first glance, this might seem to make libertarians' position boil down to much the same thing as more mainstream liberal outlooks that already treat legislation, regulation, court adjudication, and private action as complementary tools for controlling pollution. However, this interpretation of this chapter's upshot is mistaken. Even granting my arguments, libertarianism remains a distinctive position in environmental debates. Indeed, a libertarian environmentalism that makes room for legislation and regulation would still be so distinct from more mainstream liberal outlooks that we should expect many liberals to reject it strenuously.

To see the unique character of libertarians' position on pollution, let us return to the connection between compulsion and rights. Libertarians characteristically regard individuals as free to do as they please, provided they do not violate others' rights. As we have seen, this liberty implies broad latitude for carrying out private projects, entering governance arrangements, and making bargains. On the flip side, it also yields a (limited) justification for coercing those who abuse others. We have seen it is possible to use all these tools to serve environmental objectives. Crucially, however, libertarians believe it is acceptable to use compulsion *only* in response to rights-violating activities. Thus, even if libertarians grant the value of legislation and regulation for adequately protecting pollution victims, this does not imply a willingness to accept environmental legislation and regulation to serve other goals besides safeguarding rights.[38]

In the broader realm of environmental political discourse, libertarians' insistence on tethering compulsion to the protection of individual rights is idiosyncratic. Most non-libertarians believe it is appropriate to use coercive legislation and regulation to achieve environmental objectives far beyond the protection of individual rights. For instance, most consider it acceptable to use state power to protect the natural world and its inhabitants for their own sake or promote the public good through environmental projects. Non-libertarians typically do not believe societies must wait for specific rights violations before acting against those who threaten ecological integrity. Libertarianism is distinct in its refusal to countenance such interventions as a matter of principle.

Some readers may see this libertarian attitude as a liability rather than an asset to the paradigm. Insofar as libertarianism holds that we must only deploy coercion to protect individual rights—and never to promote environmental integrity or the public good more generally—this might seem so much the worse for libertarianism. I do not attempt to adjudicate this issue here. The crucial point is that this polarizing feature of libertarian environmentalism ensures the view would be distinctive even if libertarians embraced legislation and regulation more overtly. Even with its toolkit broadened beyond private action and court adjudication, libertarianism would remain a unique and likely controversial perspective in environmental debates.

Conclusion

In the context of controlling pollution, libertarians have typically disdained legislation and regulation in favor of private efforts and court adjudication. In this chapter, I have argued this attitude is misguided. Libertarians are right to see courts as a powerful tool for defending victims against pollution. However, in societies where courts operate without legislation and regulation, we should expect victims to receive inadequate protection against many common pollution problems. The key to overcoming this deficiency is to acknowledge the judiciary's circumscribed role in the liberal separation of powers and capitalize on the complementary competencies of other branches of government. By refining standards for what pollution citizens should be expected to bear, legislation and regulation can enable courts to perform the functions libertarians envision more effectively.

Whether legislators and regulators will deliver genuine improvements in practice depends on empirical factors that go beyond the scope of this chapter. In some circumstances, lawmakers and administrators will be sufficiently imperfect to make flawed court adjudication the best available option. But given the systematic limitations of the common law approach and the institutional advantages of legislation and regulation for overcoming them, we should often expect the opposite to be true. Instead of seeing legislation and regulation as "second best" alternatives to be used only as a last resort, libertarians should embrace these tools as core components of their pollution control strategy.

Notes

1. E.g., Roy E. Cordato, "Market-Based Environmentalism and the Free Market: They're Not the Same," *The Independent Review* 1, no. 3 (1997): 371–386.
2. E.g., Terry L. Anderson and Donald R. Leal, *Free Market Environmentalism for the Next Generation* (New York: Palgrave Macmillan, 2015).
3. Note that this is not to deny judges' power to adapt to novel situations. On the contrary, the common law itself is largely a product of such adaptation, as I will discuss below. The issue I highlight in this chapter has to do with the need to *change legal standards themselves* as opposed to applying them in potentially creative ways to meet new challenges. To put the issue a different way: although judges are well equipped to adapt established standards to decide new kinds of cases, they are poorly equipped within the standard liberal framework to alter established standards to decide old kinds of cases in different ways.
4. Bruce Yandle and Stuart Buck, "Bootleggers, Baptists, and the Global Warming Battle," *Harvard Environmental Law Review* 26, no. 1 (2002): 177–229; J.R. Clark and Dwight R. Lee, "Suppressing Liberty, Censoring Information, Wasting Resources, and Calling It Good for the Environment," *Social Philosophy & Policy* 26, no. 2 (2009): 272–295; A.H. Barnett and Bruce Yandle, "The End of the Externality Revolution," *Social Philosophy & Policy* 26, no. 2 (2009): 130–150; Mark Pennington, *Robust Political Economy: Classical Liberalism and the Future of Public Policy* (Northampton, MA: Edward Elgar), ch. 8; Anderson and Leal, *Free Market Environmentalism.*

5. Elizabeth Brubaker, "The Common Law and the Environment: The Canadian Experience," in *Who Owns the Environment?*, eds. Peter J. Hill and Roger E. Meiners (Lanham, MD: Rowman & Littlefield, 1998), 87–118.
6. For further discussion of the shift from principled opposition to consideration on merits, see Dan C. Shahar, "Hayek's Legacy for Environmental Political Economy," in *Interdisciplinary Studies of the Market Order: New Applications of Market Process Theory*, eds. Peter Boettke, Christopher Coyne, and Virgil Storr (Lanham, MD: Rowman & Littlefield, 2017), 87–109; Dan C. Shahar, "Environmental Issues," in *The Routledge Companion to Libertarianism*, eds. Matt Zwolinski and Benjamin Ferguson (New York: Routledge, 2022), 455–470.
7. Robert Nozick, *Anarchy, State, and Utopia* (New York: Basic Books, 1974), 28–35.
8. Morris Tannehill and Linda Tannehill, *The Market for Liberty* (Auburn, AL: Ludwig von Mises Institute, [1970] 2007), 62; see likewise Walter E. Block, "Environmentalism and Economic Freedom: The Case for Private Property Rights," *Journal of Business Ethics* 17, no. 16 (1998): 1890.
9. Peter Railton, "Locke, Stock, and Peril: Natural Property Rights, Pollution, and Risk," in *To Breathe Freely: Risk, Consent, and Air*, ed. Mary Gibson (Totowa, NJ: Rowman and Littlefield, 1985); David Friedman, *The Machinery of Freedom: Guide to Radical Capitalism* (La Salle, IL: Open Court, 1989), ch. 41; Jeffrey Friedman, "Politics or Scholarship?" *Critical Review* 6, nos. 2–3 (1992): 429–445; Mark Sagoff, "Free-Market Versus Libertarian Environmentalism," *Critical Review* 6, nos. 2–3 (1992): 211–230; David Sobel, "Backing Away from Libertarian Self-Ownership," *Ethics* 123, no. 1 (2012): 32–60; Matt Zwolinski, "The Libertarian Non-Aggression Principle," *Social Philosophy & Policy* 32, no. 2 (2016): 62–90.
10. Nozick, *Anarchy, State, and Utopia*, 57–58.
11. Eric Mack, "Elbow Room for Rights," *Oxford Studies in Political Philosophy* 1 (2015): 194–221.
12. Block, "Environmentalism and Economic Freedom," 1892.
13. Guido Calabresi and A. Douglas Melamed, "Property Rules, Liability Rules, and Inalienability: One View of the Cathedral," *Harvard Law Review* 85, no. 6 (1972): 1089–1128.

14. Carmen Pavel, "A Legal Conventionalist Approach to Pollution," *Law and Philosophy* 35, no. 4 (2016): 337–363.
15. On the role of evolved precedent in generating complex legal systems, see F.A. Hayek, *Law, Legislation, and Liberty, vol. I: Rules and Order* (Chicago: University of Chicago Press, 1973).
16. John Hasnas, "Two Theories of Environmental Regulation," *Social Philosophy & Policy* 26, no. 2 (2009): 95–129.
17. For discussion of a variety of legal strategies for tackling pollution, see Jonathan H. Adler, "Is the Common Law a Free-Market Solution to Pollution?" *Critical Review* 24, no. 1 (2012): 61–85.
18. Joel F. Brenner, "Nuisance Law and the Industrial Revolution," *The Journal of Legal Studies* 3, no. 2 (1974): 403–433.
19. *Madison v. Ducktown Sulphur, Copper & Iron Co.*, 113 Tenn. 331 (1904).
20. Ibid., 342–343.
21. Ibid., 343.
22. Ibid., 366.
23. The pagination below comes from House of Lords, *St. Helen's Smelting Co. v. Tipping, The American Law Register (1852–1891)* 14, no. 2 (1865): 104–111.
24. Ibid., 106.
25. Ibid., 109.
26. Ibid., 108.
27. Ibid., 106.
28. E.g., *Whalen v. Union Bag & Paper Co.*, 208 NY 1 (1913); *Hulbert v. California Portland Cement Co.*, 161 Cal. 239 (1911).
29. It should be stressed that finding against polluters does not have to prevent industrial activity from continuing. In cases where polluters can feasibly negotiate with would-be victims, treating victims' rights as strictly inviolable implies only that a mutually acceptable agreement must be reached before pollution proceeds. In such circumstances, pollution will only be blocked if it does not produce benefits that exceed its costs, since otherwise it will be profitable for polluters to offer ample compensation in exchange for the right to proceed. See Ronald H. Coase, "The Problem of Social Cost," *The Journal of Law & Economics* 3, no. 1 (1960): 1–44. It is because such arrangements are often costly to negotiate in practice—especially when considered at the scale of an entire industrial civilization—that it may be deemed impracticable to treat victims' interests in being free from pollution as strictly inviolable as a matter of legal principle.

30. See also on this point Peter S. Menell, "The Limits of Legal Institutions for Addressing Environmental Risks," *Journal of Economic Perspectives* 5, no. 3 (1991): 93–113.
31. E.g., B.N. Ames, L.S. Gold, and W.C. Willett, "The Causes and Prevention of Cancer," *Proceedings of the National Academy of Sciences* 92, no. 12 (1995): 5258–5265; D. Trichopoulos, D., F.P. Li, and D.J. Hunter, "What Causes Cancer?" *Scientific American* 275, no. 3 (1996): 80–87.
32. For discussion of the concept of "joint liability" as a mechanism for tackling pollution problems involving large numbers of polluters, see Adler, "Free-Market Solution," 70–72.
33. See on this point James M. Buchanan, "Politics without Romance: A Sketch of Positive Public Choice Theory and Its Normative Implications," in *The Logical Foundations of Constitutional Liberty*, 45–59 (Indianapolis, IN: Liberty Fund, [1979] 1999).
34. Montesquieu, *The Spirit of the Laws* (London: T. Evans, [1748] 1777), IX.6.
35. Catherine Sharkey, "Common Law Tort as a Transitional Regulatory Regime: A New Perspective on Climate Change Litigation", this volume.
36. For further discussion of how this might be done with an eye to libertarian theories of rights, see Shahar, "Environmental Issues."
37. Murray N. Rothbard, *Egalitarianism as a Revolt Against Nature: And Other Essays*, 2nd ed. (Auburn, AL: Ludwig von Mises Institute, [1974] 2000), 208.
38. Cordato, "Market-Based Environmentalism and the Free Market," 373–374.

Complexities of Climate Governance in Multidimensional Property Regimes

Karen Bradshaw and Monika Ehrman

Introduction

Pollution is an ancient problem with little modern resolution. Hippocrates wrote of outdoor air pollution in Greece, complaining that inhabitants "are likely to have deep, hoarse voices, because of the atmosphere, since it is usually impure and unhealthy in such places."[1] Over 2,400 years later, pollution continues to be vexing, particularly so in this time of climate crisis. Historically, environmental law regulated pollutants to protect human health. But now, the proliferation of atmospheric

K. Bradshaw (✉)
Sandra Day O'Connor College of Law, Arizona State University, Tempe, AZ, USA
e-mail: kbradshaw@asu.edu

M. Ehrman
SMU Dedman School of Law, Dallas, TX, USA
e-mail: mehrman@smu.edu

J. H. Adler (ed.), *Climate Liberalism*, Palgrave Studies in Classical Liberalism,
https://doi.org/10.1007/978-3-031-21108-9_7

and hydrologic particulate matter shifts some priority away from human health impacts due to traditional pollution sources to the direct and indirect effects related to the release of those particulates in the context of anthropogenic climate change. This chapter argues that policies and solutions are hamstrung by an incomplete framing of pollution vis-à-vis property theory, through which law continues to divide interconnected systems in ways that defy scientific understanding. By shedding light on this theoretical issue, we highlight the importance of underexplored governance tools, including localized stakeholder collaborations.

This chapter begins with a description of two important, unregulated pollutants with profound effects on human health and climate change: wildfire smoke emissions and weather modification. Regarding wildfire smoke emissions, government agencies establish wildfire policies without considering the known levels of CO_2 in smoke, creating tremendous amounts of unregulated greenhouse gas emissions, which are known to directly affect the rate of anthropogenic climate change. In weather modification, public and private entities (including for-profit companies) are allowed to release chemicals into the air to produce more rain and snow with neither regulation nor centralized recordation of how they are affecting the weather. In both instances, land management agencies, companies, and private individuals are playing fast and loose with America's health and understanding of climate change. Embracing a tech-informed, Wild West ethos of "get ahead of the regulation," unregulated actors are intentionally exposing millions of Americans to air pollutants, rendering meaningless attempts to determine human health effects or climate change effects. As the accounting truism notes, you cannot manage what you do not measure. A near-total lack of emissions accounting and regulatory oversight in these areas allows agencies and private parties to unilaterally make decisions that should be subject to democratic accountability.

This governance gap creates a troubling baseline problem. Several Environmental Protection Agency (EPA) indicators tracking trends in natural conditions—including droughts and wildfire—do not account for 50 years of prior chemical releases and landscape management policies. Without understanding what human modifications were undertaken—and how they affected key climate markers like precipitation levels and wildfire density/intensity/size—EPA indicators may inaccurately reflect climate trends over time. Our criticism does not challenge or dismiss climate science; on the contrary, we think that unregulated pollutant use has propped up climatic changes that are far more severe than existing data might suggest. Hidden action has likely long obscured the degree to which the climate has already changed.

We trace this reporting and regulatory problem to its theoretical origins in legal understandings of property. There exists an arbitrary divide between natural resources, property, and environmental law, creating theoretical silos that produce problematic real-world problems.[2] Siloed disciplines divorce the externalities of beneficial land uses from their harms, creating a legal construct that does not represent physical (natural) reality. This theoretical detachment from the physical reality of ecosystems obscures the suite of available solutions to environmental issues, leading to the hyperregulation of some pollutants and land management activities and a regulatory void around other, more pressing contributors to climate-changing activities. Governance gaps exist, with concentrated federal regulation in some areas yet little in others. Fortunately, emerging understandings of interchangeable tools to rearrange land and resource rights can address how better to match institutional arrangements to the scientific, economic, and social realities of pollution and climate change. Related governance tools—such as stakeholder collaborations operating in tandem with federal laws—have emerged to fill in the gaps.

This chapter discusses how an emerging model of multidimensional property—derived from the work of Karen Bradshaw and Dean Lueck and other emerging interdisciplinary discussions of overlapping property rights, mismatched property rights, and landscape-level resources—can improve the framing of the problem of pollution and thus the available outcomes.[3] From these observations flow new ways to assess pollution. Specifically, we argue that focusing on the human health effects of pollution without considering its effects on the environment creates intractable problems. This argument is particularly true in the wildfire context, where massive smoke emissions are obliterating climate goals in often-unaccounted-for ways in the name of the environmental benefit of restoring fire to the natural landscape.[4] Thus a multidimensional model of property provides a better understanding of the trade-offs in human influence over air, water, wind, and land.[5]

What to do? We suggest that an overreliance on federal statutory law and the concurrent underreporting of natural and anthropogenic climate modification belies the polycentric model of governance necessary for resource-based issues. This theoretical analysis leads us to suggest bolstering pollution (and therefore climate) governance through increased recognition that natural resources management requires bespoke treatment, along with scholarly recognition of the use of stakeholder groups as a conflict resolution mechanism to balance the many competing interests for natural resource property and reach socially acceptable outcomes in ways that courts, which are limited to picking winners and losers, cannot. Stakeholder collaborations serve to fill governance gaps in current environmental policy within the United States, creating a legal framework that more neatly matches the regulated issues.[6]

Climate Change Indicators

The EPA uses climate change indicators to track changes to the climate over time. The EPA justifies the use of indicators to communicate "climate science in a sound, transparent, and easy-to-understandable way."[7] The indicators track seemingly neutral metrics and show changes over time, which is attributed to climate change.

This chapter argues that EPA's reliance on climate change indicators is premature, as the indicators do not account for the associated legal and regulatory effects. Just as humans have long managed the landscape, so too have we long managed the airspace—intentionally and effectively affecting rainfall, snowpack, and wildfires. These anthropogenic effects of modifying ostensibly natural phenomena are largely unnoticed, unregulated, and thus unconsidered by climate accounting. This lack of regulation and accounting renders climate trends unreliable. In essence, the problem with using these indicators is that they do not account for human influence on these measures over time.

Critically, this paper is not one of climate skepticism. As stated above, our questions surrounding the veracity of indicators do not challenge climate change. On the contrary, it is impossible to know in which direction the missing data cuts. Accurately accounting for presently uncounted policy inputs may show that the climate has changed far more than we presently realize. Human action to control the physical environment may have obscured for decades the extent to which the climate would have otherwise already changed.

Weather Modification

The EPA uses weather and climate indicators that measure rainfall, snowfall, and the timing of snowfall. The EPA indicator for the US and global precipitation patterns are based on measurements, beginning in 1901, from land-based weather stations.[8] Importantly, these measurements do not account for human-caused weather modification, such as cloud seeding.

Cloud seeding is an artificial weather modification currently used to increase precipitation from certain cloud systems for drought mitigation, agricultural, and recreational (e.g., increasing snowpack for ski hills) purposes.[9] It involves the release of silver iodide or dry ice from airplanes and weather towers into clouds to cause rainfall and snowfall. Scientists developed cloud seeding technology in the 1950s. It was used in American conflicts with Vietnam and Laos as a form of meteorological warfare and later became the subject of a UN Charter, restricting weather modification for peaceful purposes. Although the theory of generating precipitation by seeding clouds was generally accepted, data supporting any successful efforts was rare, and is only recently forthcoming.

Cloud seeding occurs when seeding material in aerosolized form is released into certain cloud systems. In the US, cloud seeding is governed by a single federal law, which allows both public and private entities and individuals to engage in this unregulated practice after first registering with the National Oceanic and Atmospheric Administration (NOAA). Although registration does not involve oversight of cloud seeding, it affords NOAA the opportunity to monitor records maintained by registrants.

The absence of regulatory oversight has not impeded cloud seeding activity. Since the 1970s, states, public utilities, and companies have used cloud seeding to increase snowpack and spring and summer runoff for municipal and agricultural water supplies. For example, in a 2013 report, Utah noted it had engaged in cloud seeding since 1973, which increased runoff by 59.2 billion gallons. The state estimated that cloud seeding created a 10.9% increase in precipitation during 2012–2013. Although weather modification occurs, there is no centralized record of modification information, including who is doing it and the effects of modification on precipitation.[10] Scattershot pieces of information provide some insights; but incorporating such data appears impossible because it does not appear to exist.[11]

In addition to muddying EPA climate indicators, the use of cloud seeding may have still-to-be-determined detrimental impacts on the environment and human health. Although users of silver iodide for seeding state that the silver levels in the resulting precipitation are below EPA secondary maximum contaminant levels,[12] there simply is not enough research or data to conclude that there is no or little effect to the dispersion of silver iodide.[13] And as with any incineration or combustion process, amounts of carbon dioxide or other non-combusted products may result. Though these amounts may be de minimis (really, they are unknown), an increase in cloud seeding activity over decades may produce cumulative harmful impacts on environmental wellbeing. The absence of regulatory oversight also appears to include routine individual project or programmatic environmental assessment, as required under the National Environmental Policy Act. Although private and other nonfederal actors may be conducting the weather modification activity, applications are submitted to NOAA.

Meanwhile, the technical reports of the EPA indicators and the underlying scientific papers contain no mention of the effects of weather modification on indicators, including snowpack and drought. For example, one technical report notes that variability can include "observation methods, land use, and forest canopy" but do not contemplate human-caused sources.[14] To put it simply, if we have been making rain, it is unsurprising that it is raining more. Counting precipitation levels without accounting for human-created weather modification seemingly creates a faulty comparison of precipitation levels over time. As a result, efforts to attribute changes in precipitation levels to climate change without considering the effects of intentional weather modification paint an inaccurate picture of anthropogenic climate change.

Wildfire

The EPA also considers wildfire as a climate change indicator. "This indicator tracks wildfire frequency, total burned acreage, burn severity, and the seasonality of wildfire activity in the United States from 1983 to 2020."[15]

Wildfire suppression policy has changed dramatically in the period underlying the indicator. For many decades, federal government firefighting policies followed the "10 am rule," which required firefighters to extinguish each fire prior to 10 a.m. the following morning.[16] This total suppression policy focused on putting out fires completely and quickly. As a result, wildfire sizes remained small; they were extinguished quickly; and they burned relatively little area. In the 1990s and early 2000s, fire ecologists began questioning the wisdom of suppression. A narrative of restoring fire to the natural landscape emerged. Critiques of suppression claimed that this policy increased fuel load, resulting in hotter, more intense, more damaging fires. Over time, agencies formally changed wildfire policy, shifting from total suppression to "let burn" policies.[17] These policies are essentially the opposite of the "10 am rule," allowing fires to grow and burn without suppression attempts. Unsurprisingly, letting a fire burn unimpeded results in larger, longer, and more intense fires.

Fire suppression policy underwent a dramatic about-face from the 1980s to the present. Yet, continuing to use the fire indicator without accounting for this policy shift results in a potential false attribution. Simultaneously, land management policies on federal public lands also significantly shifted during this time. Timber harvest dramatically fell on public lands, while pest infestations created diseased trees more susceptible to wildfire. This shift was not accounted for in the wildfire policy and thus existing climate science.

Because we are not accounting for these meaningful policy changes, we cannot accurately gauge whether and how much climate change has increased fire risk. The EPA indicator technical documentation notes that wildfires can be influenced by factors other than climate, "particularly changes in fire suppression and fire management practices … [and] resources available to fight and manage wildfires …"[18] The report notes, "further analysis is needed before an apparent change in wildfire activity can necessarily be attributed to climate change."[19] So although countless examples of political statements and media reports attribute the intensity, duration, and destructiveness of wildfire to climate change, these accounts crucially overlook policy changes to public land management and fire suppression policies.

Worsened wildfires have multidimensional effects on issues of environmental justice, greenhouse gas emissions, and human health effects from smoke emissions. For example, wildfire smoke emissions have grown so large that California wildfires emit as much carbon dioxide as a year of electricity use.[20] Against the pressing considerations of climate change effects like sea-level rise affecting low-lying island nations and coastal populations, comprehensive policy analysis should weigh the beneficial effects of restoring wildfire to the natural landscape against the harms the United States creates for other countries through unremedied smoke emissions. There exist many open, unanswered scientific questions that should inform these policy considerations. But without climate science that accounts for climate change indicator policy changes, much is left on the table.

The remainder of this chapter overviews how a new approach to land management and pollution can inform legal and regulatory analysis. Adopting integrated pieces of the larger, interconnected system might produce new solutions to understanding and addressing pressing climate change issues.

Siloed Approaches in Interconnected Systems

The foundational elements of earth, fire, air, and water constantly change as part of a dynamic and interconnected system. Any change to one affects the others. But law and policy generally contemplate human interference with these elements individually. For example, the Clean Air Act considers air quality and the Clean Water Act governs water. Although this individualistic governance appears logical and actionable, it is not sufficiently attuned to the genuine way in which complex natural systems operate.[21] These elements do not exist in neatly contained silos, but instead interact and influence one another in respects that scientists do not still understand and, most certainly, current environmental laws do not fully contemplate. As research and technology improve and globalization increases, scientists increasingly are finding examples of teleconnection—where apparent anomalies are, in fact, related. But because of this current isolation, federal laws fail to reflect the consequences of changes to essential elements of the larger, interconnected systems.

Siloed thinking within environmental law hampers efforts to address observable issues. One can readily observe that the elements have changed throughout our lifetime. Just as Garrett Hardin observed that water quality varied between his grandfather's lifetime and his own,[22] so too are many people observing that those places they intimately know are noticeably different than they once were—hotter, drier, stormier, more polluted, etc. These observations of natural conditions are essentially human—we notice the local effect of human influence on earth, fire, air, and water, but are very poor at recognizing the causes. Siloed disciplines and their resultant homogenous knowledge may also prevent us from seeing solutions.[23] EPA's climate accounting suffers from the same causal oversights. Namely, it observes changes in natural conditions but often fails to account for their human causes. Climate science divorced from regulatory and policy inputs may be failing to account for human effects on the climate.

We argue that the conceptualization of the discrete areas of natural resources, property, and environmental law creates arbitrary divisions that divorce the externalities of beneficial land uses from their harms.

These divisions create artificial legal constructs that do not represent physical reality. This intentional theoretical detachment from physical (and observable) reality obscures the suite of available solutions to environmental issues and prevents the identification of additional problems. Fortunately, emerging understandings of interchangeable tools to rearrange land and resource rights can address how better to match institutional arrangements to the scientific, economic, and social realities of pollution and climate change.

From these observations emerge new processes to assess pollution. Specifically, we argue that focusing on the human health effects of pollution without considering its other environmental effects (positive and negative) creates intractable problems.

Unregulated Polluters

We often talk about "landscape management"—the concept that humans actively manage the land by planting, fertilization, tilling, harvesting, etc. In this context, property law generally envisions individual landowners managing their land as they please, within the confines set by law. We less often discuss "water management," "atmospheric management," or "firescape management." Yet, public and private parties directly control water, atmosphere, and fire, just as they manage lands. Given the surprising lack of regulation for non-land governance zones, individual humans can create profound effects that affect human health, property value, and the environment of related systems. Through the release of chemicals into one system, we (in)directly affect another. A homeowner using rat poison can decimate a food chain; chemicals in the water turn whales into toxic, polluting sources[24]; emissions from the combustion of petroleum in cars produce acid rain, falling over land and ocean.

Pollution is a complicated problem. In most cases, it is not the substance itself that is inherently hazardous to human health such that it must be regulated, but rather the amount of the substance. Toxicology teaches that substances become injurious in small amounts but hazardous in others. One ice cream cone is not all that bad; five might make you sick; ten a day over a lifetime will contribute to many health conditions.

Pollution is similar: a little pollution may be ok, but prolonged pollution exposure at a heightened rate creates detrimental effects on human health.

Most pollutants that pose serious public health risks come from a variety of sources. Famously, methane emissions that contribute to climate change come not only from cars on the road and airplanes in the sky but also the belching of cattle.[25] Governing pollution, then, means sifting through emitting sources and deciding which among them may emit how much. This picking of winners and losers is complex and localized, creating the cooperative federalism model of the Clean Air Act, in which EPA sets the allowable limits of criterion pollutants and states are left with the task of determining who might pollute how much. Alternative mechanisms of allocation include markets, such as carbon, which are premised on the theory that the highest-value user will purchase credits from the lower-value user.

One unspoken aspect of pollution regulation is unregulated emission sources. Some actors are subject to considerable, costly regulation, while others escape wholly unregulated.[26] In this way, regulated polluters and affected populations bear the costs of unregulated polluters. Regulatory absence forces these groups to either reduce their pollution levels or increase health risks because the government fails to reduce levels of unregulated pollution.

Pollution's Connection to Land

When we think of pollution, we focus on air and water—not earth (land) or fire. Yet, many laws historically governing pollution derived from English property laws, which were developed to manage land and resources. The puzzle of why landlocked rules might cover pollutive air or water emissions may seem archaic; but in fact, regulated pollution largely derives as an externality of production. And production turns a location-based natural resource into a product with human utility. Because the natural resource is tied to the land, pollution therefore begins as a land-based problem.

To analyze this long-obscured link, consider lamb fleece becoming wool, cotton plant fibers becoming textiles, or timber becoming lumber. The industrial process of transforming raw agricultural products into usable goods produces pollutants. These pollutant chemicals occur in physical production, air emissions, and effluent. Economic constraints govern how far the harvested good can travel prior to being processed, with transportation costs to centralized processing facilities (mills) being a limiting factor on the viability of production. In other words, the pollution emitted from processing natural objects into usable goods necessarily must take place within a predictable radius of the place from which the goods are grown and harvested.

Consider textiles: textiles may begin as cotton, which is grown on the land, processed, and turned into usable products. Pollutants may begin with the agricultural runoff of chemically treated fields; be contributed to by the trucks that carry the unprocessed cotton to the mill; seep into the air and water through the mill processing facility; and again occur through truck transportation.

This localizing effect of industrial pollution may appear to be a limiting principle upon its production. Why would a city allow pollution where its children play outside, subject to the known detrimental health effects of high concentrations of harmful emissions on human health?[27]

Moreover, the broader, systemic issues of economic and racial injustice result in historically marginalized communities shouldering a disproportionate pollution burden.[28] Environmental justice concerns abound. Minority and low-income populations are disproportionately located near polluting facilities,[29] which is of heightened concern as new research indicates associations between pollution and health issues, including diabetes, asthma, and cancer.[30]

A Theory of Landscape-Level Governance

Referring to weather and fire as natural resources may seem peculiar; but they share the same geophysical governance challenges of other landscape-level resources, such as water basins, forests, wildlife habitats, and grazing ranges. Firesheds—landscape-scale areas that face a wildfire threat—operate at a regional scale—smaller than federal resources, larger than localized resources. Fire does not follow jurisdictional boundaries; nor does it respect property boundaries. It encompasses a great deal of land comprised of a mix of underlying landowners: individuals, corporations, states, tribes, and the federal government. Effective policy therefore requires cooperation.

Firesheds also contain highly variable geographies and resource mixes, such as homes, trees, wildfires, livestock, and watersheds. Population density, resource access, and land values also vary widely. Laws designed to protect hillside homes in California differ from those that should manage grassfires in Texas or tundra fires in Alaska.

Natural resource governance is complicated because it must encompass multiple jurisdictions and property classifications—e.g., federal and local, public and private. For these reasons, we believe that Congress's historic hands-off approach to wildfire was correct. Congress tasked land management agencies with administering public lands and then left them alone to do so. Agencies subsequently created highly decentralized systems in which the career progression of employees was dependent upon satisfying supervisors and residents. This balance was complicated. Sometimes problems resulted in litigation, but mainly federal land management agencies handled problems at a local level.

During a fire, the law consolidates firefighting authority into a single source—the government. The emergency doctrine clearly hands unilateral decision-making power to the government.[31] Efficient disaster response requires a single party making decisions to mitigate harm during the disaster, rather than compensating parties afterward.

Coordinating fire prevention efforts is far more difficult. In nonemergency situations, property rights generally trump environmental regulation. Landowners possess diverse preferences for cutting trees, creating defensible space, and building homes in wildland-urban interface areas. In this regard, maximizing the value of an individual's property may conflict with sensible landscape-level management.[32]

Absent government condemnation of fire-prone private property, ex ante mitigation of wildfire risk requires landowners to voluntarily cooperate to create landscape-level risk management structures. In most Western states, this cooperation involves public–private arrangements and may include multiple owners and managers because of historical land policies. Thus, land within the same fireshed can be managed by individual homeowners, states, tribes, and federal agencies. Wildfire management is paradoxically both highly localized and necessarily federal—whether the fire burns on public or private land.

Wildfire mitigation and ex post restoration strategies are highly localized. Yet, federal land management agencies alone have the resources, expertise, and land management authorities to implement solutions. About 30% of the United States is composed of lands managed by federal agencies. Because the federal government is the largest landowner in the United States, it follows that federal land management agencies are best positioned to influence mitigation policy. But the often administratively unwieldy size and over-expansive agenda of the federal government are not desirable characteristics of a land manager. For example, the White Mountain Apache Tribe has employed successful wildfire mitigation strategies. In 2011, the Wallow Fire burned about half a million acres, which is still Arizona's worst fire. Without the Tribe's forest management practices, which mimic the natural burn-and-growth cycle, the fire would have been much more severe.[33]

> Reflecting on the stark contrast between tribal and federally managed lands, Jonathan Brooks, the tribal forest manager, hinted as to why. "The forests for the White Mountain Apache Tribe, they're very important for livelihood, for economics, cultural aspects, recreation. There's so many benefits that the land and that the forests provide for the tribe, and it's very important for us to actively manage it to keep the forest healthy so that everything kind of maintains its balance," he said. Key to that success is the tribe's being "unhindered by environmental litigation and drawn-out federal government processes," he added.[34]

Successful wildfire solutions, therefore, must be a hybrid governance strategy, both federal-local and public–private in nature.

Stakeholder Collaborations

Stakeholder groups, which manage landscape-level resources, may provide a long-overlooked piece of the solution to pollution and managing non-land resources (such as watersheds and airsheds). Stakeholder collaborations allow iterative, responsive policymaking with flexibility unavailable through other means—a crucial tool in responding to ever-changing land and resource conditions.

Interestingly, federal land management agencies have largely privileged local concerns since inception. Long before Elinor Ostrom won a Nobel Prize for describing the benefits of resource users engaging in the governance of natural resources,[35] agencies like the Forest Service, Bureau of Land Management, and National Park Service were crafting innovative partnerships with stakeholders to retain the benefits of federal control governed with local knowledge.

Over the past several decades, a local-first view of federal land management became anathema to environmental activists. Through federal policies, environmentalists were able to achieve tremendously valuable gains in vital issues like air and water pollution. Yet, federal laws centered on natural resources—such as the Endangered Species Act and National Environmental Policy Act—are less clear successes. Policymakers did then what they are dangerously poised to do now: fail to understand that natural resources cannot be governed under a primarily top-down, federalist model. Any environmental problem that has a land-based component cannot be resolved through federal law alone, but instead requires polycentric governance. One-size-fits-all federal policies are problematic and prone to failure.

In the natural resource context, hierarchical federal law certainly achieved valuable outcomes from an environmental preservation perspective. But they also created a tremendous backlash among rural populations and landowners. Federalized policies can overlook important localized knowledge, such as the traditional ecological knowledge of Indigenous groups. "While many of the changes associated with increasing state and market integration have established more resilient infrastructures in some regions of the world, they have also frequently undermined traditional adaptations of rural populations to natural hazards."[36]

Win–win solutions are, in some ways, less satisfying because they require compromise, unlike win-loss outcomes. There is less to advertise to constituents and fewer clear-cut gains. Pragmatically, small wins may be the most realistic. Achieving these outcomes is entirely possible. Indeed, over a dozen federal land management agencies have over a thousand stakeholder groups managing public lands and natural resources. But, until now, this invaluable tool for capturing the benefits of federal public land and the need for localized resource governance has been wholly unnoticed by legal scholars.[37] It exists in the interstices of administrative law and the study of natural resources, a seldom-explored area that is only recently emerging.[38]

Weather modification and wildfire policy operate within blind spots within administrative and environmental law. Administrative law scholars tend to focus on the Environmental Protection Agency, disregarding the thirteen administrative agencies that govern public lands and natural resources (water, wildlife, forests, etc.) But EPA primarily focuses on pollution, largely from industrial emitters, with command and control regulation. Natural resource governance is different from pollution in a critical, fundamental way.[39] Scientific and sociological understandings suggest that land-based problems operate within ever-changing socio-ecological structures with complex biological interactions. They are subject to pressures not present to the governance of resources untethered from geographic space.

The foundational misunderstanding is that the "land" should be treated separately from the ecological system. Unlike a first-year Property class, the reality of true natural resources management disallows a discrete parceling of rights and sticks or neat severance of surface from the subsurface. All living and nonliving organisms interact within the physical environment. Modern agency practice and the above-mentioned academic study overlook this integrated dynamic. Additionally, there remains general ignorance of past deliberate resource management efforts by Indigenous populations and wildlife, which have interacted and managed natural resources for tens of thousands of years and through geologic time respectively.

Negative-Value Resources

In introducing "negative-value property" to the property law lexicon, Bruce Huber described land with so much toxic material that the cost of remediating is more than the value of the underlying property.[40] In this section, we extend Huber's concept to natural resources, noting that there are many natural features (some of which are not even called resources) that run adjacent to the land and resources in physical space. Where we might talk about oil and gas reserves, airspace, and water as resources that run with the land, we are less likely to talk about pest areas, firesheds, predators, and fuel load.[41] But maximizing the commercial value of some resources on the land (timber and cattle, for example) requires suppressing other negative-value resources (wildfire and pests).

Putting the land to "beneficial use" or "best and highest use" requires a landowner to decide which resource to maximize and then allocate resources differently. These are subjective decisions. A forester might choose to suppress wildfire to maximize timber harvest. A conservationist might choose to allow a wildfire to restore the natural landscape. A cattle rancher might choose to kill predators to maximize beef values. A wildlife advocate might introduce wolves onto the landscape to increase the population. At times, differing land management policies sharply collide. Gates of the Arctic National Park in Alaska cultivated a wolf population, building on the values of conservation and preservation. Alaskan state officials waited in helicopters outside the park and gunned down the entire wolf pack as it crossed the federal-state border, prioritizing the values of hunting caribou over wolf conservation. Negative-value attributes receive less attention.

This analysis shows how divisions between land management and natural resources (such as restoring fire to the natural landscape to benefit forest health) and the pollution control focus of environmental law create gaps between the cause of pollution and its effects. It follows that analysis of risk management developed in the pollution context may not extend to the context of the natural resources. This conclusion renders conventional risk management strategies inappropriate in land management and natural resource contexts. Thus, scholarly analysis of risk management that does not specifically inform land management agencies cannot be readily applied to wildfire risk reduction. Although most scholars suggest that the EPA should use cost–benefit analysis to govern pollution, it does not follow that the Forest Service should not use the same risk management tool to govern wildfire.

Wildfire, A Case Study

We use wildfire as an example of how an integrated analysis of pollution reveals factors obscured from a siloed view considering only atmospheric emissions. Fire itself is a natural process, at once destructive and regenerative. Yet it undeniably causes devastating effects, with dozens killed and hundreds of thousands of people made homeless. Wildfire is an environmental justice disaster, with rural, socioeconomically disadvantaged, and Indigenous persons disproportionally affected. Years after a wildfire burned Paradise, California, vulnerable residents still live in camping tents in the parking lot of a nearby Walmart.[42]

Although it is less visible, the emissions from wildfires are also devastating to vulnerable communities worldwide. In 2018, the wildfire emissions in California obviated all of the CO_2 emission reductions the state had obtained through other measures. As outlined above, state and federal agencies managing wildfire emissions are unregulated polluters. This section considers how we might consider that through the lens of multidimensional property.

For the first two hundred years of American history, the federal government did not provide a comprehensive policy for wildfire.[43] In his excellent, comprehensive overview of wildfire law, Robert Keiter noted that "the most striking fact about fire management on the public lands was the relative dearth of law on it."[44] Federal agencies that manage public lands had the authority to develop wildfire policies.[45] Agencies, in turn, were highly responsive to state and local considerations.[46]

Congress began "fixing" wildfire in 2013. Since then, the incidence of wildfire has dramatically increased. What if our new approach to wildfire is causing the crisis? Indeed, the sharp uptick in disastrous wildfires correlates with increasing federal control and an increased reliance on data-driven fire modeling. When other national disasters occur, we discuss commonly understood solutions. When it comes to wildfire, there is no soundbite solution. Wildfire is inherently local: what works for an Alaskan tundra fire may fail for a Texas prairie wildfire. Yet, human choices always affect the size, duration, direction, and intensity, of a wildfire. The local interests most affected by wildfire should drive the policies that manage it.

Everyone respects firefighters who perform heroic work under extreme conditions. But the agency governing them is rife with perverse incentives. When the conservation movement pushed the Forest Service to stop cutting down trees, it pivoted towards fighting fire. For decades, the Forest Service had a blank check for wildfire spending—it could outspend its budget, then ask Congress for reimbursement. Even today, the agency pays employees dramatically increased overtime the longer a fire burns; premises career advancement, in part, on active fire days; and omits wildfire smoke from its emissions limits. Sovereign immunity prevents courts from second-guessing agency firefighting decisions. Under these conditions, the Forest Service has dramatically shifted from fighting fires to managing them—moving away from immediate suppression.

Historically, government firefighters attempted to suppress wildfires as quickly as possible. Federal government firefighting efforts effectively subsidized private forestry. Gifford Pinchot's model of renewable forestry is only commercially viable with wildfire suppression. Foresters transformed many natural forests with large, widely spaced trees into densely packed, monoculture tree farms, which ecologists argue increased the risk and intensity of wildfire.

In the 1970s, environmentalists and fire scientists began questioning the wisdom of suppressing fires. They noted that fires have occurred in forests since time immemorial; they are an integral part of the natural landscape. Plants like the Lodge Pole Pine rely on fire to open their cones and reproduce. Big wildfires also serve less direct ecological goals. President Obama once stated that fires are hotter and larger than they were in the past as evidence of climate change, without accounting for agencies' dramatically changed wildfire suppression and forest management strategies. Larger wildfires decrease the likelihood that people will build homes in forests, limiting urban sprawl. And the viability of commercial timber operations depends on reduced wildfire risk. Letting wildfires burn has far broader socio-ecological effects than opening pinecones.

People who build homes in forests have an ever-increasing stake in wildfire management. Local municipalities grow their property tax base by allowing homeowners to build in desirable natural areas known as the wildland-urban interface areas. These homes increase the likelihood that humans will be at risk during a wildfire, which in turn prompts firefighters to engage in riskier techniques to save lives. Homes also distract from protecting timber and, by extension, watersheds, and wildlife. Insurance laws in California mandate that homes in forests—just like those on coastlines—must be insurable, providing a bar on the market response that might otherwise keep people from building there.

Are the terrible losses in California attributable to policy choices made to satisfy a romantic ideal of a landscape that no longer exists? Or, is the real problem government subsidies putting commercial forestry and wildland-urban interface users in a natural environment where they do not belong? These questions are too complicated and nuanced to be distilled into a soundbite or tweet. National policy cannot resolve these conflicts. Empowering the stakeholders who bear the costs and benefits of the forests to generate better approaches may be our best step forward.

Restoring Fire to the Natural Landscape

The conversation around restoring wildfire to the national landscape contains some objectionable premises. What, exactly, does "natural" mean? If it implies pre-Colonial landscape management practices, that blindly ignores Indigenous peoples and their widespread historical landscape management practices (which sometimes included intentional prescribed burning).[47] History aside, it is difficult to imagine most North American forests today as "natural." Commercial timber harvest depends upon a process of cutting trees, replanting seedlings, allowing them to grow, then planting them again.

Foresters replant seedlings designed to grow quickly to maximize revenue. As a result, the timber stands in many forests are monocultures or comprised of a limited mix of commercially valuable trees. As such, the wildfire regime appropriate in a "natural" context is distinctly inappropriate for forests that are no longer "natural." To imagine the difference between a natural forest and a modern forest that was subject to decades of commercial timber practices, imagine a raw swath of wild prairie and a cornfield. In one, there are several plants. In the second, even-age management of selected seeds of the same plant variety grows. To suggest that a landscape management practice suited toward the wild prairie would be appropriate to treat the cornfields without accounting for the differences between the underlying landscapes would be poor advice. Similarly, historic wildfire practices applied to modern landscapes without updating to accommodate the reality of commercially managed lands does not make sense.

Similarly, the "natural" model of forests on which the restoration of wildfire is advocated does not include houses. Real forests, however, increasingly do. The expansion of homes into traditional-forested areas—sometimes called the wildlife urban interface ("WUI") area—represents the largest risk of restoring fire to the natural landscape. Although the public might be complacent as forestry diminishes, few are comfortable watching homes burn. The WUI of homes in forests can include wealthy landowners. And it is here that federal firefighting agencies have drawn the line—quite literally—with firefighting. Although federal firefighters might allow fire in sparsely populated areas, they are statutorily required to protect human life and homes.

In non-obvious ways, wildfire mitigation and management directly impact whether forests are used for commercial timber production or left in their natural states. For decades, federal agencies managed risk as private foresters continue to—by cutting trees. Agencies also followed stringent rules to suppress fire as quickly as possible. Over time, however, agencies reflected an increased environmental focus. When goals extended beyond timber harvest, ecologists argued that there were benefits to the plant and animal life in an ecosystem accrued by wildfire. Who can argue with science?

Yet, ostensibly neutral science has a normative edge. This argument implicitly challenges the domestic forestry industry—without adequate wildfire protection, commercial forestry is not viable in the United States. In this sense, supporting wildfire as a natural part of the ecosystem was a mechanism for promoting conservation. Certainly, forest landowners have pushed back—hard—against federal policies to allow fires to burn, understanding that such policies affect their economic interests.

With respect to reform, solutions fall into one of three categories: (1) ex ante land management policies to prevent or mitigate harms created by wildfire; (2) fire suppression strategies to save human lives and satisfy other objectives while the fire is burning; and (3) ex post fire recovery rules to lessen the negative effects, such as mudslides.

"Managing" wildfire risk requires controlling human action on a natural physical landscape. Consider a few examples: insurance companies requiring homeowners to replace wooden shingles with metal roofs to reduce the likelihood that a burning ember will light a house on fire; local governments advocating for defensible space around a home, clearing brush, and other flammable vegetation near a home; and foresters suggesting that cutting trees reduces the fuel load, which reduces the incidence and severity of wildfires.

A mix of groups governs wildfire risk. Insurers' and timber landowners' coordinated policies represent private governance. Local, state, and federal governments manage adjacent lands. Yet, wildfire does not respect property lines. It operates at a much larger scale than an individual parcel of land. One neighbor's risk management strategy affects the others. Thus, any strategy must be cooperative.

The sensible way to manage a landscape is to group ownership at varying levels to reflect different scales of efficient management. Defense, airspace, and radio waves are most sensibly organized at the national level. Multiple layers of governance simultaneously exist on, above, under, and appurtenant to the same piece of land.[48]

This overlapping, multiple-owner land system is complicated. Some argue that divestiture of federal land results in the best use and prioritization of rights, while others contemplate government-owned resources as a representative democratic form of ownership. There is no "correct" option—conflicts abound on federal land, as they do on fee. The reality of overlapping property ownership and management is likely a best case, especially if ownership composition is not a binary choice (i.e., individual or government). Professor Jonathan Adler theorizes that large landowners could also be composed of trusts and various other collective entities, which simultaneously encompass a broad value spectrum while retaining key characteristics of private property ownership.[49] Other public–private and cooperative partnerships may also be more useful structures for subnational resources, such as water and wildfire, which cannot be managed efficiently at either a federal or individual level. Encouraging structured decision-making with landowners can encourage effective management, while balancing multiple objectives, including use, development, conservation, and recreation. Federal decision-making with the inputs of stakeholders directly interacting with the resource may provide a path forward.

Conclusion

In this chapter, we add to emergent literature by analyzing how natural resource property contains a multitude of rights, interests, and systems and the institutional mechanisms devoted to managing such complexity. We conclude that this intricate problem demands a variety of institutional tools, which may change in response to shifts in social values, resource dynamics, or economic conditions over time. This analysis yields a new way of looking at pollution and climate change through the lens of multifaceted property.[50]

Acknowledgements The authors thank Jonathan Adler, William Boyd, Gregory Wolcott, and the participants of the IHS Workshop on Liberty, Property, and Pollution and members of the George Mason Law and Economics Workshop for their helpful comments. Challie Facemire and Sarah Brunswick provided outstanding research assistance.

Notes

1. David Fowler, Peter Brimblecombe, John Burrows, Mathew R. Heal, Peringe Grennfelt, David S. Stevenson, Alan Jowett, Eiko Nemitz, Mhairi Coyle, Xuejun Liu, Yunhua Chang, Gary W. Fuller, Mark A.

Sutton, Zbigniew Klimont, Mike H. Unsworth, and Massimo Vieno, "A Chronology of Global Air Quality." *Philosophical Transactions of the Royal Society A*, 378 (2020): 1–28, https://doi.org/10.1098/rsta.2019.0314.

2. We extend the idea of hidden resources, introduced by Monika Ehrman in other recent work, to the issue of climate-based pollution to illustrate a hyper focus on landscape management and neglected understanding of human management of airscapes, watersheds, and firescapes; Monika Ehrman, "Application of Natural Resources Property Theory to Hidden Resources," *International Journal of the Commons* 14, no. 1 (2020): 627–637, http://doi.org/10.5334/ijc.982; Monika U. Ehrman, "Hidden Resources," *UC Irvine Law Review* 13 (forthcoming 2023).
3. Karen Bradshaw, Billy Christmas, and Dean Lueck, "An Introduction to 'Overlapping Resources and Mismatched Property Rights,'" *International Journal of the Commons* 14, no. 1 (2020): 553–556, http://doi.org/10.5334/ijc.1070.
4. William Boyd, "Climate Liability for Wildfire Emissions from Federal Forests," *Ecology Law Quaterly* 48, no. 101 (2022): 101–134, https://doi.org/10.15779/Z38BN9X40F.
5. Carol M. Rose, "Thinking about the Commons," *International Journal of the Commons* 14, no. 1 (2020): 557–566, http://doi.org/10.5334/ijc.987.
6. Challie Facemire and Karen Bradshaw, "Biodiversity Loss, Viewed through the Lens of Mismatched Property Rights," *International Journal of the Commons* 14, no. 1 (2020): 650–661, http://doi.org/10.5334/ijc.985.
7. "Frequent Questions about Climate Change Indicators," *U.S. Environmental Protection Agency*, last modified July 18, 2021, https://www.epa.gov/climate-indicators/frequent-questions-about-climate-change-indicators.
8. Except for Alaska measurements, which began in 1925; "Climate Change Indicators: U.S. and Global Precipitation," *U.S. Environmental Protection Agency*, last modified July 17, 2021, https://www.epa.gov/climate-indicators/climate-change-indicators-us-and-global-precipitation.
9. Cloud seeding may also be used as a form of geoengineering for climate change mitigation, whereby precipitation decreases associated rising temperatures and drought.
10. We advance a novel observation that the nexus of weather modification and the advent of the relatively new, $10 billion crop insurance

industry merits suspicion and further studies. Although this book chapter provides only a preliminary treatment of the topic, we see nothing in the laws regarding weather modification that would preclude insurers from either directly engaging in cloud seeding to cause rain or being alerted in advance of cloud seeding by third parties. Given that crop insurance policy payouts are related to precipitation levels, there appears to be an open question as to whether insurance companies could—or are—using weather modification to manage payouts.

11. An attempt to gain this information from the National Oceanic and Air through a Freedom of Information Act request produced no response.
12. "Secondary Drinking Water Standards: Guidance for Nuisance Chemicals," *U.S. Environmental Protection Agency*, last updated January 19, 2021, https://19january2021snapshot.epa.gov/sdwa/secondary-drinking-water-standards-guidance-nuisance-chemicals_.html.
13. Utah Department of Health, https://health.utah.gov/enviroepi/appletree/technicalassists/Silver_Iodide_Cloud_Seeding.pdf.
14. U.S. Environmental Protection Agency, "Technical Documentation: Snowpack," *U.S. Environmental Protection Agency*, 2021, 7, https://www.epa.gov/sites/default/files/2021-04/documents/snowpack_td.pdf.
15. U.S. Environmental Protection Agency, "Technical Documentation: Wildfires," *U.S. Environmental Protection Agency*, 2021, 1, https://www.epa.gov/sites/default/files/2021-04/documents/wildfires_td.pdf.
16. Karen Bradshaw, "Norms of Wildfire Suppression Among Public and Private Landowners," in *Wildfire Policy: Law & Economics Perspectives*, eds. Dean Lueck and Karen M. Bradshaw (New York: Routledge, 2012), 89.
17. Karen Bradshaw, "A Modern Overview of Wildfire Law," *Fordham Environmental Law Review* 21, no. 3 (2010): 445–478; Karen Bradshaw, "Backfired! Distorted Incentives in Wildfire Suppression," *Utah Environmental Law Review* 31, no. 1 (2011): 155–180.
18. U.S. Environmental Protection Agency, "Norms of Wildfire Suppression Among Public and Private Landowners," at 11.
19. Ibid.
20. U.S. Department of the Interior, "New Analysis Shows 2018 California Wildfires Emitted As Much Carbon Dioxide As an Entire Year's Worth of Electricity," *U.S. Department of the Interior Press Release*, November 30, 2018, https://www.doi.gov/pressreleases/new-analysis-shows-2018-california-wildfires-emitted-much-carbon-dioxide-entire-years.

21. Karen Bradshaw and Bryan Leonard, "Virtual Parceling," *International Journal of the Commons* 14, no. 1 (2020): 597–610, https://www.jstor.org/stable/27016587.
22. Garrett Hardin, "The Tragedy of the Commons: The Population Problem Has No Technical Solution; It Requires a Fundamental Extension in Morality," *Science* 162, no. 3859 (1968): 1245, https://www.jstor.org/stable/1724745.
23. Jonathan Adler (Case Western Reserve University School of Law Professor) IHS and Case Western Workshop comments, October 16 and 24, 2021.
24. Sarah Zielinski, "Whales Are Full of Toxic Chemicals," Wild Things: Animals, Science News, January 19, 2016, https://www.sciencenews.org/blog/wild-things/whales-are-full-toxic-chemicals#:~:text=Persistent%20organic%20pollutants%20are%20not%20even%20the%20only,also%20works%20its%20way%20into%20whales%20and%20dolphins.
25. Amy Quinton, "Cows and Climate Change: Making Cattle more Sustainable," Food: News, UC Davis, June 27, 2019, https://www.ucdavis.edu/food/news/making-cattle-more-sustainable.
26. Union Electric Co. v. Environmental Protection Agency, 427 U.S. 246 (1976). The Court found that state implementation plans under the Clean Air Act are not subject to findings of economic or technological feasibility requirements by the Environmental Protection Agency.
27. The choice of "children" here is not incidental. Children—because of their developing bodies and small size—are particularly vulnerable to human health effects from high level of exposure to certain pollutants. Childhood exposure to pollutions can lead to lifelong health problems.
28. For example, see Julia Jacobo's article "Communities with Large Black, Asian and Hispanic or Latino Populations more likely To Be Exposed to Air Pollution, New Research Suggests," ABC News, https://abcnews.go.com/US/communities-large-black-asian-hispanic-latino-populations-exposed/story?id=82199331.
29. Robert V. Percival, Christopher H. Schroeder, Alan S. Miller, and James P. Leape, *Environmental Regulation: Law, Science, and Policy*, 8th ed. (Philadelphia: Wolters Kluwer, 2018): 20 (Citing United Church of Christ Justice and Witness Ministries, *Toxic Wastes and Race at Twenty: 1987–2007* [2007]).
30. Ikenna C. Eze et al., "Association Between Ambient Air Pollution and Diabetes Mellitus in Europe and North America: Systematic Review and

Meta-Analysis," *Environmental Health Perspectives* 123, no. 5 (2015): 381–389, https://doi.org/10.1289/ehp.1307823.
31. Thomas Merrill, "Property and Fire," in *Wildfire Policy: Law & Economics Perspectives*, eds. Dean Lueck and Karen M. Bradshaw (New York: Routledge, 2012): 32–49.
32. Robin Kundis Craig and J.B. Ruhl, "Adaptive Management for Ecosystem Services across the Wildland-Urban Interface," *International Journal of the Commons* 14, no. 1 (2020): 611–626, http://doi.org/10.5334/ijc.986.
33. Katie Tubb, "How Federalism Is Making a Difference on Western Lands," Environment: Commentary, The Heritage Foundation, August 21, 2019, https://www.heritage.org/environment/commentary/how-federalism-making-difference-western-lands.
34. Ibid.
35. Elinor Ostrom, *Governing the Commons: The Evolution of Institutions for Collective Action*, 1st ed. (Cambridge: Cambridge University Press, 2009).
36. Gregory Button and Anthony Oliver-Smith, "Disaster, Displacement, and Employment: Distortion of Labor Markets During Post-Katrina Reconstruction," in *Capitalizing on Catastrophe, Neoliberal Strategies in Disaster Reconstruction*, eds. Nandini Gunewardena and Mark Schuller (Lanham: AltaMira Press, 2008): 123–147.
37. Certainly, scholars have widely recognized the use of public-private partnerships in other areas of Administrative Law. But, such partnerships are generally cast as new. In the natural resources contexts, they have been operating for hundreds of years because collaboration is the only way to effectively manage a landscape.
38. *See* Karen Bradshaw, "Agency Engagement with Stakeholder Collaborations, in Wildfire Policy and Beyond," *Arizona State Law Journal* 51, no. 2 (2019): 437–504; Karen Bradshaw, "Stakeholder Collaboration as an Alternative to Cost-Benefit Analysis," *Brigham Young University Law Review* 2019, no. 3 (2020): 655–724; Karen Bradshaw, "Stakeholder Dynamics in Land Development," *The Journal of Legal Studies* 50, no. S2 (2021): S53–S70, https://doi.org/10.1086/705921.
39. One way to read Elinor Ostrom's prize-winning is that natural resource governance differs from pollution control. Legal scholars have yet to link that to Administrative Law, but we do so here.
40. Bruce R. Huber, "Negative-Value Property," *Washington University Law Review* 98, no. 5 (2021): 1461–1515.

41. Tara Kathleen Righetti, "Liberating Split Estates," *International Journal of the Commons* 14, no. 1 (2020): 638–649.
42. Dani Anguiano, "Anxious Wildlife Refugees Camped in Parking Lot Wonder: Where Now?" *US News, The Guardian*, November 19, 2018, https://www.theguardian.com/us-news/2018/nov/19/wildfire-refugees-california-chico-walmart-parking-lot-latest.
43. Readers interested in the historical development of wildfire in the United States should review the work of historian Stephen Pyne, the preeminent wildfire historian in the United States. He has written countless books and articles on the topic. For a good general overview, *see* Stephen J. Pyne, *Fire: A Brief History*, 2nd ed. (USA: University of Washington Press, 2019).
44. Robert B. Keiter, "The Law of Fire: Reshaping Public Land Policy in an Era of Ecology and Litigation," *Environmental Law* 36, no. 2 (2006): 322.
45. Ibid.
46. Karen Bradshaw, "New Governance and Industry Culture," *Notre Dame Law Review* 88, no. 1 (2013): 2515–2550.
47. *See generally*: Roxanne Dunbar Ortiz, *An Indigenous Peoples' History of the United States* (Boston: Beacon Press, 2014).
48. Bradshaw and Leonard, "Virtual Parceling."
49. Jonathan H. Adler, "Taking Property Rights Seriously: The Case of Climate Change," Social Philosophy and Policy 26, no. 2 (2009): 296–316; Jonathan H. Adler, "Nature and the Firm," this volume.
50. Reproduced from "Unregulated Climate Mitigation", in Vol. 35 of *Georgetown Environmental Law Review* (forthcoming 2023) by Karen Bradshaw & Monika U. Ehrman with permission from Karen Bradshaw & Monika U. Ehrman.

Climate Change and Class Actions

Brian T. Fitzpatrick

In this chapter, I explore the possibility of using class action lawsuits to generate solutions to the problem of climate change. The lawsuits I envision might be, for example, based on the simple common law of nuisance. One difficulty with using individual nuisance lawsuits to combat climate change is that there are so many parties affected by climate change that the number of potential litigants is basically limitless. This precludes private ordering: if an emitter strikes a deal with one litigant, nothing forecloses another litigant from suing over the same emission. As Jonathan Adler put it, "no plaintiff (and, indeed, no defendant) will forego litigation in pursuit of a cooperative solution as there is no mechanism to prevent suit by another party."[1] It is often thought that only government regulation can cut through the transaction costs

B. T. Fitzpatrick (✉)
Vanderbilt Law School, Nashville, TN, USA
e-mail: brian.fitzpatrick@vanderbilt.edu

J. H. Adler (ed.), *Climate Liberalism*, Palgrave Studies in Classical Liberalism,
https://doi.org/10.1007/978-3-031-21108-9_8

posed by widespread externalities. But here I explore whether the class action lawsuit might offer another way.

What Is a Class Action?

In a class action lawsuit, one or more litigants can represent a class of all similarly situated litigants. The class members do not participate in the litigation; they are passive. But whatever happens in the lawsuit will bind them and they will not be able to sue later on their own. If the representative loses, the rights of all the class members are extinguished as well. If the representative wins, the other class members win, too; if the representative wins money, he or she splits the money with everyone else. Unlike other countries, in the United States, affirmative consent to this arrangement is not required, and, in many class actions, not even notice to the class is required; rather, the court is responsible for scrutinizing the representative to ensure the representative is typical of the rest of the class and adequate to the task.[2] The representative will, of course, hire a lawyer to litigate the case and the court will scrutinize the law firm as well to ensure it will adequately represent the class. But all of the work is performed by the law firm and the representative. For this reason, the marginal cost of adding another member to a class is basically zero. This is what makes the class action an alternative solution to the private ordering problem discussed above.

One might ask why we allow other people to litigate for us like this in the United States. The answer is that, in this country, litigation has always been the primary means of regulating behavior. As I explain below, we use litigation much more than administrative regulation to deter misconduct and to compensate people when misconduct has nonetheless occurred. Yet, sometimes misconduct will harm us in such small amounts that individual litigation is not practical. How many thousands, tens of thousands, or perhaps even hundreds of thousands of dollars would you need to be harmed before you would hire a lawyer? How big would the case need to be before the lawyer will agree to take it? Class actions are a means to solve the transaction-cost problem of aggregating all of these small, individual harms into a litigation vehicle that will be worthwhile for a lawyer to pursue. Because aggregation with no marginal cost can become very lucrative very quickly, class action lawsuits have attracted the attention of entrepreneurial lawyers who figure out how to aggerate and present claims that would otherwise go unaddressed.

If all of this sounds well and good, I should note that the lack of affirmative consent in American class actions is a serious issue for classical liberals. Letting other people do things to us without our authorization is not one of our hobbies. But, even in America, putative class members have a right to "opt out" of the class in most cases. This opt-out right is arguably a second-best form of consent. Moreover, most class actions seek relief that is too small for individuals to pursue on their own so it is unclear what functional benefit there is to requiring consent here. In addition, in order to solve the private ordering problem discussed above, widespread class membership is necessary. If class members had to "opt in" to be included in a class action as they do in most other countries, class membership would not be widespread given the difficulties of notifying all putative class members and the inertia they would have to overcome to join the lawsuit when their individual stake in the lawsuit is small.

A class action can be brought on behalf of a class against one or more defendants (known as a "plaintiff class"), or a plaintiff can bring a class action against a class of defendants (known as a "defendant class"). Defendant classes are very rare because they pose a number of practical obstacles.[3] Plaintiff classes are common, but rare in the environmental field.[4] This may be because plaintiffs typically seek injunctive relief rather than damages in environmental cases and it is often unnecessary to bring a class action to obtain complete injunctive relief: an individual plaintiff can win a court order stopping an emission just as well as a class of plaintiffs can. For the reasons stated above, that is not the case with climate change litigation: class action lawsuits are all but necessary to obtain complete relief. Moreover, it is possible that climate change class actions could seek damages as well. If they do, they could be very lucrative. The most lucrative class action lawsuits in American history were both environmental cases: the multi-billion dollar class action against BP for the Deepwater Horizon explosion and oil spill off the coast of Louisiana and the multi-billion dollar class action lawsuit against Volkswagen for its diesel engine emissions.[5]

My focus in this chapter will not be the doctrinal questions that climate change class actions might pose; there are many legal hurdles to certifying class actions. My focus will be on whether class action lawsuits might be preferable to government regulation as a policy matter. To put the question another way: would it be better to have parties negotiating in the shadow of class action lawsuits or in the shadow of government regulations?

For a host of reasons I describe below, I think private class action litigation is often the better way to intervene in the market. I make many of these arguments in a recent book called *The Conservative Case for Class Actions*.[6] Although the title uses the word "conservative" to appeal to a broader audience, the arguments in the book are mostly from the perspective of a classical liberal and that is the terminology I will use here. But given the perspective from which I wrote the book, the case I will make here will obviously not be for everyone.

I should note that I am not certain the case I make in the book applies perfectly to climate change litigation. As I will explain, much of the case turns on the comparative advantage of the private bar over the government with respect to motive—i.e., the purity of the profit motive over the impurity of ideological or special interest capture. But the profit motive requires lawyers working for contingency fees and pursuing damages on behalf of their clients; it is not clear to me to what extent climate change class actions would seek damages rather than injunctive relief. If they are not seeking damages, then lawyers who bring these suits might be dominated by public interest lawyers who are the very definition of special interests.

But let me turn to why I think class action lawsuits are usually preferable to government regulation; along the way, I will discuss why these arguments might or might not apply to climate change class actions.

The Classical Liberal Case for Class Actions

To see the case for class action lawsuits, it is helpful to start by stepping back and examining the alternatives for legal intervention in a systematic way. There are four basic means by which we might bring legal interventions to bear on voluntary market transactions. I map them in Fig. 1. On one axis, we see a choice between enforcing the law before someone like an emitter acts (ex ante) or after someone like an emitter acts (ex post). This is the choice between requiring permission before you act versus being permitted to do whatever you want (but having to pay up later if things don't turn out well). On the other axis, we see a choice between who does the enforcement: the government or a private party.[7]

Most developed countries around the world intervene in the market through box 1: you have to ask permission before you do something new, and you go to the government for that permission. If the government tells you that what you want to do is lawful, then you are good to go. These countries deal with any harm that results through social insurance programs.

The United States mostly falls into box 4: you don't have to ask anyone's permission, but you have to pay for any harm you cause, and the mechanism to collect those payments is initiated by whoever is injured. We don't need social insurance programs to pick up the tab; compensation comes from lawsuits.

Most proposed climate change regulation falls into box 1. Private class action litigation aggregating small harms against climate change emitters would fall into box 4. The choice between boxes 1 and 4 should not be difficult for a classical liberal. Many have said as much in the past.[8] They have included academics like Milton Freidman[9] as well as politicians like the libertarian Republican Gary Johnson.[10] Indeed, a terrific book by a classical liberal research fellow at George Mason University's Mercatus

	Ex Ante	Ex Post
Government	1	2
Private	3	4

Fig. 1 Enforcement choices

Center is devoted entirely to this question; its conclusion could not be clearer: "*ex post* (or after the fact) solutions should generally trump *ex ante* (preemptive) controls."[11] The reason is simple: box 1 stifles innovation and experimentation because we can never know enough about something new to know how much permission to grant to it. It also requires massive taxes to support social insurance programs to compensate people when they are harmed. Box 4 lets people innovate and experiment as much as they want as long as they promise to clean up any messes they make. And it costs us nothing in taxes.

Progressive economists make the case for box 1 by arguing that judges have neither the incentives nor the expertise to fashion rules of liability for market behavior. As I explain below when comparing boxes 2 and 4, I am very skeptical that decentralized, independent, generalist judges are inferior to centralized, politically-compromised, albeit specialized, government bureaucrats.[12] The case is even more dubious for box 1. As Andrei Shleifer explains, "[w]ith respect to the creation of rules, there are even deeper concerns about regulators than about judges."[13] Indeed, although comparative studies of this sort are difficult to do well, we now have empirical evidence that box-4 nations have better economies than box-1 nations. For example, Rafael La Porta, Florencio Lopez-de-Silanes, and Andrei Shleifer have recounted studies showing "the superior performance of... common law countries."[14]

But what if the government climate change intervention took place in box 2? Some commentators have proposed combating climate change through government *parens patriae* lawsuits for nuisance. This would be another way of pursuing small harms in the aggregate. Would private class action lawsuits still be preferable to the government if the government was acting *ex post*? Both box 2 and box 4 take advantage of the innovation and energy that comes from letting people do what they want without asking for permission first. Both of these boxes seek to deter wrongdoing by giving people incentives to be careful about what they do by insisting that you pay for any harm you cause later on. Neither of these boxes requires the creation of social insurance programs to compensate people when the permitted activities injure people; the actors themselves pay the compensation when they are sued later on.

The only difference is who brings the lawsuits: government lawyers or private lawyers.

So which lawyers should classical liberals prefer? I think the answer is usually pretty easy here, too: private lawyers. Indeed, there was a time when this notion would not have been as provocative as it might sound today. Although it has been largely forgotten, for most of our history, those sympathetic with classical liberalism preferred legal enforcement by private lawyers because they thought private enforcers of the law were better than public enforcers. For example, in the 1970s, the prominent economists Judge Richard Posner, William Landes, Gary Becker, and George Stigler engaged in a famous debate on this question. In a series of articles, they debated who is better suited to enforce the criminal and civil law: private parties or the government?[15] Becker and Stigler said it was private parties largely because they have better incentives to detect misconduct and pursue enforcement: "[private] enforcement is good precisely because the incentives to enforcers are as large as the incentives to prospective violators."[16] I develop this argument in more detail below. Posner and Landes worried that letting *anyone* sue to enforce violations of law—i.e., even if they were not the victim of the misconduct—would lead to wastefully duplicative efforts to detect misconduct, but, when the misconduct is easier to detect, even they favored private enforcement.[17] Other classical liberal thinkers in this era came to the same conclusion. For example, Friedrich Hayek famously endorsed the common law over legislation not just because of the virtues of judicial over political lawmaking, but also because private enforcement was a form of spontaneous ordering.[18]

It was not just in the academy that classical liberals had these thoughts. They manifested themselves in the political world as well. As Robert Kagan[19] and Sean Farhang[20] have chronicled, many of the statutory regimes Congress enacted in this era could win Republican support only on the promise that they would be enforced by private lawsuits rather than government bureaucrats. Indeed, for much of the twentieth century, it was progressives who objected to private lawsuits to enforce the law. One of the reasons progressives built the administrative state during the Progressive and New Deal eras was to wrest enforcement of the law away from the private sector.[21] Franklin Delano Roosevelt went so far as to veto New Deal legislation when it relied too heavily on private enforcement instead of government agencies.[22] Similarly, decades later, it was the Carter Administration that sought legislation to abolish small-claim class actions brought by the private bar and replace them with government lawsuits.[23] The sponsor of the Administration's bill in the United States Senate? Ted Kennedy.[24]

In my book, I try to reclaim this classical liberal tradition, but I do so by linking it to the theory of privatization of government. Since at least the 1970s, the theory of privatization has been a central tenet of classical liberal theories of government. For many of the same reasons we want to privatize nearly everything else, I argue we should want to privatize the enforcement of civil laws.

The theory of privatization is often traced to Margaret Thatcher's British government in the late 1970s, but Robert Poole, the founder of the Reason Foundation (a leading privatization think tank) is said to have coined the term in the 1960s.[25] Whatever its origin, it has been a staple of Republican politics and classical liberal thought in the United States since Ronald Reagan.[26] The basic idea is that much of what the government does should be done by the private sector. The theory encompasses a spectrum of efforts to transition government work to private parties.[27] At one end, the government entirely divests itself of assets or industries, as Britain did with many of its industries under Thatcher and as many want the United States to do with Amtrak.[28] On the other end, more common in the United States, the government retains financial control but outsources the delivery of goods or services to private parties.[29] There are numerous arrangements in between. There is almost no end to the government services that classical liberals want to privatize in one form or another.[30]

Why do we love to privatize? In the book, I identify six reasons I found in the literature: (1) smaller government is better than bigger; (2) self help is preferable to dependence on government; (3) private actors have better incentives; (4) private actors have access to better resources; (5) private actors are more independent from special interests; and (6) private actors are less centralized.[31] As I show there, all six of these reasons counsel in favor of using private lawyers over government lawyers to enforce the law.[32] In this chapter, I have space to elaborate on only the two most powerful reasons: better incentives and more independence.

Better Incentives

There is little question that the profit motive usually gives the private bar better incentives to enforce legal rules than those of government lawyers. Government lawyers, like all government employees, generally earn the same salary no matter how much action they take against wrongdoers. They also enjoy the same civil service protection from termination as many government employees. Whether they bring one lawsuit or 10, whether they win or lose, they still have a job and they still make the same salary. This is not so for the private bar. Most class action lawyers work on "contingency": they only get paid if they win and the more they win, the more they get paid. Scholars from all walks of political life think that the contingency fee gives the private bar a leg up over government lawyers,[33] including scholars affiliated with the Chicago School of economics.[34] Indeed, even the most vocal critics of the so-called "private attorney general" concede this point. Consider Walter Olson, who authored the famous anti-lawsuit book *The Litigation Explosion*: "There is no point denying that contingency fees have certain productivity advantages. Paying people only if their efforts culminate in success definitely coaxes more effort out of them."[35]

It is true that government lawyers can be motivated by political ideology and that this can induce them to do a good job even without financial rewards. But it can also induce them to do a bad job. Ideological lawyers might enforce only the laws they like and not the ones they don't. It is another species of the bias that affects government enforcement that I discuss below.

Most class action lawyers, by contrast, are motivated by money and not ideology. But will that be true of climate change class action lawyers? I am not sure. On the one hand, many environmental lawsuits are brought not for damages but for injunctive relief; that blunts (but does not eliminate) the profit motive. Moreover, many of the lawyers bringing environmental lawsuits today work for public interest organizations that are very ideological. On the other hand, in theory, there is no reason climate change class actions cannot seek damages like other environmental lawsuits have—and when these lawsuits have sought damages in the past, they have attracted the very best of the profit-motivated private bar. As I noted above, the most lucrative class actions in American history have been environmental cases.

Ironically, one of the biggest complaints about class action lawyers from right-leaning circles is that they are so profit motivated. Some believe that enforcement of the law is different from other products or services that might be produced in the private sector because the profit motive will always drive lawyers to take things too far and file too many lawsuits. The concern is that profits can be made not only pursuing egregious misconduct; profits can also be made pursuing conduct that is neither egregious nor even in violation of the law. I understand this concern. It makes sense that the pursuit of profits leads the private bar to exploit technicalities, to push the envelope on what is illegal, and to file meritless lawsuits. It also makes sense that, because government bureaucrats cannot pursue profits and have more limited resources, they do these things less often.[36] On the other hand, turning enforcement over to the government is not the only way we can inject more discretion into the enforcement of the law. In order for private plaintiffs to win lawsuits, they must convince a judge to interpret the law in their favor; if judges think the lawsuits are nitpicky technicalities not worthy of the court's time, they can dismiss them.

Moreover, nothing about this concern is unique to lawyers' profit motives. It is a well-known problem of the profit motive that, if not pointed in the right direction, it can drive people to do bad things.[37] Many progressives complain about corporate profit motives for these same reasons. Corporate profit motives can lead corporations to cut corners when they make products, to deceive customers about what they are buying, and to conspire with their competitors to fix prices. As good classical liberals, our response to these problems is not, as it has been in other countries, to nationalize all of our industries. Our response is to acknowledge that profit motives can lead to both good and bad, and to put laws into place that point corporate motives more toward the good than the bad.

Our answer should be the same when it comes to profit-motivated lawyers. Profit-motivated lawyers are no different than profit-motivated anything else. Because they are profit motivated, they will enforce the law more thoroughly than government lawyers will. This means they will bring more lawsuits against corporate misconduct. But it also means that, if we let them, they will bring more lawsuits that we are not so keen on. A rising tide lifts all lawsuits, so to speak. What we have to do is not cast the private lawyer aside, but to guide their incentives with rules that point their lawsuits in the right direction—just as we sometimes have to guide the corporate profit motive through various rules of the marketplace.[38]

It is true that some are pessimistic that we can guide lawyer profit motives well enough. The concern is that it is hard to calibrate lawyers' fee awards so that we achieve the socially optimal level of enforcement or that the legislature or judges are not up to the task. But the same is true of every profit motive, including the corporate ones. It is hard to calibrate the rules of the market to ensure corporate motives are pointed in the right direction and they have big lobbying budgets to try to resist regulation. But we would rather try our best than to turn our industries over to the government. Dare I say, the same is true of the enforcement of legal rules.

More Independence

Government enforcers are beset by the distractions of special interest campaign money, lobbying, and the revolving door just as much as other government officials. Indeed, government enforcers may be the government officials *most* affected by this crony capitalism. Businesses have every incentive to influence the government to look the other way when they do something wrong or to give them a sweetheart deal if it doesn't look the other way. The government has enormous discretion in deciding when to enforce the law and when not to, in deciding when to settle a case and when not to. Who's to say the decision wasn't made on the merits as opposed to past election support? The promise or even hope of future election support? The fact that the wrongdoer is run by a former colleague and friend from the government? No one.

Scholars agree: government enforcers are often "captured" by the businesses and industries against which they are supposed to be enforcing the law.[39] Indeed, it is classical liberal scholars who are often the *most* agitated about government capture—hence, again, our focus in recent years on crony capitalism. As one scholar notes, "Libertarians and conservatives have been particularly critical of the progressive state because of its propensity to special interest capture."[40] Some trace this entire field of inquiry to the classical liberal Chicago[41] and Virginia Schools of economics.[42] Again, we would expect capture to lead to fewer enforcement actions and lower recoveries even when an enforcement action is brought.

As I noted above, it is an open question in my mind whether the lawyers who would bring climate change class actions would be free from special interest capture. If the lawyers are the same ones who bring much of the environmental litigation today, the answer to that question would be no: these lawyers work for public interest organizations that are the very definition of special interests. But, as I suggested above, if climate change class actions pursue damages, then I suspect they will attract the same profit-motivated members of the private bar who bring other big stakes environmental litigation. Those lawyers are more likely to keep their eyes on the ball than the government is.

It is true that the private bar cannot bring wrongdoers to account all on their own; they have to bring their cases in court. But aren't courts run by judges, and aren't judges government officials? Can't the judges be captured by businesses and industries like other government officials? If so, what difference does it make whether a private lawyer or a government one brings the lawsuit?

It makes a big difference in federal court where most climate change class actions would be litigated. Federal judges are not like other government officials. They never have to run in an election and they do not work for people who have to run in elections. Unlike other public officials in America (and, indeed, almost the entire world), federal judges can keep their jobs for life if they want to. Short of the improbable act of impeachment, a federal judge cannot be threatened. This was the entire reason the founding generation of our country conferred such job protection on them; they were obsessed with guaranteeing federal judges "independence." Moreover, the staff of a federal judge—so-called law clerks—turns over frequently (every year or two) and it is extraordinarily rare for someone to leave the staff for industry and come back again. That is, the doors of a federal judge's chambers do not revolve; they slam shut as you leave. For these reasons, few people think they can be captured like other government officials.[43]

* * *

As I said, these are only two of the reasons private litigation is usually preferable to government litigation, but, as I explain in the book, all of the other reasons classical liberals prefer private solutions line up in favor of private litigation as well. Thus, to the extent class action litigation might be able to solve the private ordering difficulty presented by the externalities of climate change, class action litigation might well be worth trying before turning to the government.

Notes

1. Jonathan Adler, "Stand or Deliver: Citizen Suits, Standing, and Environmental Protection," *Duke Environmental Law & Policy Forum* 40, no. 1 (2001): 79.
2. For divergent approaches in other parts of the world, see generally Brian T. Fitzpatrick and Randall Thomas, eds., *The Cambridge Handbook on Class Actions: An International Survey* (United States: Cambridge University Press, 2021).
3. I found only three among 688 class action settlements in federal court in 2006 and 2007. Brian T. Fitzpatrick, "An Empirical Study of Class Actions and Their Fee Awards," *Journal of Empirical Legal Studies* 7, no. 4 (2010): 817–818.
4. They are so few that I listed them in the "other" category (which altogether made up only 6% of the total) when I canvassed the federal class action settlements discussed in the previous note. See ibid. at 818 (notes to tbl. 1).
5. I discuss these cases in Brian T. Fitzpatrick, *The Conservative Case for Class Actions* (Chicago and London: University of Chicago Press, 2019), 82–83.

6. See ibid.
7. For a description of some of the virtues and vices of each of these boxes, see Daniel Kessler, "Introduction," in *Regulation versus Litigation: Perspectives from Economics and Law*, ed. Daniel Kessler (Chicago and London: University of Chicago Press, 2011), 13–22.
8. Herbert J. Hovenkamp writes that "[l]ibertarians and conservatives have been particularly critical of the progressive state ... [in] contrast ... [to] the common law". "Appraising the Progressive State," *Iowa Law Review* 102, no. 3 (2017): 1086–1087.
9. See Milton Friedman and Rose Friedman, *Free to Choose: A Personal Statement* (San Diego, CA: Harcourt Brace Jovanovich, 1990), 207.
10. See Ryan Lizza, "The Libertarians' Secret Weapon," *The New Yorker*, July 25, 2016.
11. Andrew Thierer, *Permissionless Innovation: The Continuing Case for Comprehensive Technological Freedom* (Arlington, VA: Mercatus Center, 2014). See also Veronique de Rugy, "Beyond Permissionless Innovation," *Reason*, January 2016, 14.
12. See, for example, Andrei Shleifer, "Efficient Regulation," in *Regulation versus Litigation: Perspectives from Economics and Law*, ed. Daniel Kessler (Chicago and London: University of Chicago Press, 2011), 31–42.
13. Ibid., 39. See also Steven Shavell, "A Fundamental Enforcement Cost Advantage of the Negligence Rule over Regulation," *Journal of Legal Studies* 42, no. 2 (2013): 275.
14. "The Economic Consequences of Legal Origins," *Journal of Economic Literature* 46, no. 2 (2008): 286, 298.
15. See Gary S. Becker and George J. Stigler, "Law Enforcement, Malfeasance, and Compensation of Enforcers," *Journal of Legal Studies* 3, no. 1 (January 1974): 1–18; William M. Landes and Richard A. Posner, "The Private Enforcement of Law." *Journal of Legal Studies* 4, no. 1 (January 1975): 1–46.
16. Becker and Stigler, "Law Enforcement, Malfeasance, and Compensation for Enforcers," 13.
17. See ibid., 3, 30, 32.
18. See Peter J. Boettke and Rosolino Candela, "Hayek, Leoni, and Law as a Fifth Factor of Production," *Atlantic Economic Journal* 42, no. 2 (2014): 129–130.
19. See Robert Kagan, "American Adversarial Legalism in the Early 21st Century," (unpublished manuscript, March 2015), 50–51.

20. See Sean Farhang, *The Litigation State: Public Regulation and Private Lawsuits in the U.S.* (Princeton, NJ: Princeton University Press, 2010), 107–108, 120.
21. See Andrei Shleifer, *The Failure of Judges and the Rise of Regulators* (Cambridge, MA: MIT University Press, 2012), 143, 148. See also William B. Rubenstein, "On What a Private Attorney General Is—and Why It Matters," *Vanderbilt Law Review* 57, no. 1 (2004): 2135.
22. See Harry Kalven Jr. and Maurice Rosenfield "The Contemporary Function of the Class Suit," *University of Chicago Law Review* 8, no. 4 (1941): 686. The text of the veto message is available online through a historical archive maintained by the University of California, Santa Barbara at http://www.presidency.ucsb.edu/ws/index.php?pid=15914.
23. See Stephen B. Burbank and Sean Farhang, *Rights and Retrenchment* (Cambridge University Press, 2017), 12.
24. See ibid.
25. Alfred C. Aman Jr., *Government by Contract: Outsourcing and American Democracy*, eds. Jody Freeman and Martha Minow (Cambridge, MA: Harvard University Press, 2009), 262. See also the Reason Foundation's citation of the 1969 book *The Age of Discontinuity* by Peter Drucker and the 1970 book *Uncle Sam, the Monopoly Man* by William Wooldridge for conceiving of privatization as an intellectual movement. Reason Foundation, *Annual Privitization Report 2006: Transforming Government Through Privitization* (Los Angeles: Reason Foundation, 2006), 21.
26. See *Transforming Government*, 2–36.
27. See, e.g., John D. Donahue, *The Privitization Decision:Public Ends, Private Means* (New York: Basic, 1991), 7. E. S. Savas lists this spectrum of activity from more privatized to less privatized: market, franchise, vouchers, grants, contract, government vending, intergovernmental agreement, and governmental. *Privatization and Public–Private Partnerships* (New York: Chatham House Publishing, 2000), 88, Table 4.6.
28. See Donahue, *The Privatization Decision*, 215.
29. See ibid.
30. Fitzpatrick, *The Conservative Case for Class Actions*, 30–31.
31. Ibid., at 31–33.
32. Ibid., at 34–44. Many of these points were made from a less ideological perspective in an important article many years ago, Samuel Issacharoff, "Group Litigation of Consumer Claims: Lessons from the U.S. Experience," *Texas International Law Journal* 34, no. 1 (1999): 135.

33. David Freeman Engstrom cites scholars who argue "profit-motivated private enforcers supply salutary forms of legal innovation". "Private Enforcement's Pathways: Lessons from Qui Tam Litigation," *Columbia Law Review* 114, no. 8 (2014): 1918. Engstrom goes on to note that a "common claim is that entrepreneurial private enforcers serve as incubators, bringing fresh ideas to the regulatory marketplace" ibid. at 1931. Engstrom writes that "champions assert that the profit-motivated nature of private enforcement renders it more cost-effective and nimbler than poorly incentivized or sclerotic public enforcement bureaucracies" ibid. at 1925. John C. Coffee Jr. theorizes that "the private attorney general is induced by the profit motive to seek out cases that otherwise might go undetected" (220), or to "intensif[y] the penalty [in cases that are otherwise detected]" (223). "Rescuing the Private Attorney General: Why the Model of the Lawyer as Bounty Hunter is Not Working," *Maryland Law Review* 42, no. 2 (1983): 215–288. In another article, Coffee reasons that "the plaintiff's attorney in these actions behaves as a risk-taking entrepreneur" "Understanding the Plaintiff's Attorney: The Implications of Economic Theory for Private Enforcement of Law Through Class and Derivative Actions," *Columbia Law Review* 86 (1986): 677. Margaret H. Lemos posits that "government attorneys ... have 'low powered' incentives compared to private employees" (quotation marks omitted). "Privatizing Public Litigation," *Georgetown Law Journal* 104, no. 3 (2016): 540. Lemos expands on this in another article, commenting that "[t]he percentage-fee arrangement under which most private class counsel work has the advantage of linking the interests of attorney and clients: the attorney's fee increases as the clients' recovery grows." "Aggregate Litigation Goes Public: Representative Suits by State Attorneys General," *Harvard Law Review* 126, no. 2 (2012): 515. Anthony Casey and Anthony Niblett present evidence "that the court-centric private-plaintiff [qui tam] mechanism is superior to the agency-centric mechanism when there is asymmetric information". "Noise Reduction: The Screening Value of Qui Tam," *Washington University Law Review* 91, no. 5 (2014): 1174–1175. Alexandra Lahav observes that "the interests of the lawyers bringing these cases are well-aligned with the merits of the legal action. Many private lawyers are paid only if they win". *In Praise of Litigation* (Oxford: Oxford University Press, 2017), 38–39. Martin Redish observes that "[t]he concept of the bounty hunter holds a venerable position in our nation's history .... [T]he pursuit of private gain motivates private individuals to expose legal activity, thereby

supposedly furthering the broader public interest in having the corporate world adhere to the broad behavioral proscriptions set by governmental authorities." *Wholesale Justice: Constitutional Democracy and the Problem of the Class Action Lawsuit* (Palo Alto, CA: Stanford University Press, 2009), 26. Later, Redish comments that "[p]rivate litigation may often do the government's work for it, by deterring and punishing violations of law. By seeking to benefit the individual litigants ... adjudication [has] the incidental impact of advancing the public interest" (31). Lemos, with Professor Max Minzner, note that some of the fire of the profit motive can be introduced among government lawyers if the departments they work for are permitted to keep the enforcement monies they recover instead of remitting them to the general treasury. See Lemos and Minzner's chronicling of "when enforcement agencies are permitted to retain all or some of the proceeds of enforcement" and how they accordingly "are likely to behave more like private enforcers than is commonly appreciated" even though "their employees are paid by salary" and do not profit financially. "For-Profit Public Enforcement," *Harvard Law Review* 127, no. 3 (2014): 523.

34. Becker and Stigler write that "[t]he essence of victim enforcement is compensation of enforcers on performance ... instead of by a straight salary." "Law Enforcement," 14. They expand upon this point, theorizing that

> the right of amount of self-protection by potential victims is encouraged, not the excessive (wasteful) self-protection that results when victims are not completely compensated, or the inadequate self-protection that results when they are automatically compensated. Further, the rewards of innovation will spur technical progress in private enforcement as in other economic callings. (15)

35. Walter, Olson, *The Litigation Explosion: What Happened When America Unleashed the Lawsuit* (New York: Penguin, 1991), 42.
36. See Richard A. Nagareda, "Class Actions in the Administrative State: Kalven and Rosenfield Revisited," *University of Chicago Law Review* 75, no. 2 (2008): 615–618. But see Engstrom's counterpoint that "one might expect a similar trend in regimes delegating enforcement authority solely to prosecutors and agencies." "Private Enforcement's Pathways," 1936.
37. See, e.g., Olson, *The Litigation Explosion*, 42.

38. See, e.g., Coffee, *Entrepreneurial Litigation*, 221. Compare this view with A. Mitchell Polinsky's stance which, although critical of private enforcement, concedes that "[r]egulating private enforcers by paying them something different than the fine for each violation detected can achieve the socially most preferred outcome in the competitive case". "Private versus Public Enforcement of Fines," *Journal of Legal Studies* 9, no. 1 (1980): 108.
39. Peter H. Schuck notes that "[t]here are many reasons for limited, weak enforcement, ... [including that] the officials who run the regulatory agencies tend to be mindful of the need—both political and bureaucratic—to cultivate and sustain the regulated industry". *Why Government Fails So Often* (Princeton University Press, 2014): 221–222. Schuck goes on to explain that the "enforcement ... and administrative implementation of regulatory statutes is always shaped by a process in which industry lobbying plays an important role" (223). Furthermore, Schuck observes that "politicians sometimes seek to restrain enforcement against putative violators who are their constituents, allies, or favored interests" (226). Coffee argues that "*elected* state officials are particularly likely to serve as unfaithful agents of class members' interests." *Entrepreneurial Litigation*, 192, emphasis in original. Engstrom observes that "public prosecutors and agency administrators have their own mix of politically inflected and careerist incentives that can cause public enforcement to deviate from the ideal." "Private Enforcement's Pathways," 1930. Engstrom goes on to note that agencies are "imperfect guardians" because of "regulatory 'capture'" and "politically conscious agencies will allocate resources with an eye to collecting political rewards and ensuring the continued flow of resources to the agency" (1939). Ramphal recalls that "for most of its history the [SEC] has been plagued by ... the political whims of all varieties of politicians." Nishal Ray Ramphal, "The Role of Public and Private Litigation in the Enforcement of Securities Laws in the United States," PhD diss., Rand Graduate School: 2007, 103. However, Lahav notes that "private litigants are ... not subject to regulatory capture, as government employees looking toward their next job in the private sector might be." *In Praise of Litigation*, 38. Zackary D. Clopton suggests that "public enforcers may be less likely to pursue cases against their political allies." "Class Actions and Executive Power," *New York University Law Review* 92, no. 4 (2017): 893.
40. Hovenkamp, "Appraising the Progressive State," 1086.

41. See Shleifer's citation of "the Chicago School" for the proposition that "the political process of regulation is typically captured by the regulated industry itself." *The Failure of Judges*, 4.
42. See James M. Buchanan and Gordon Tullock, *The Calculus of Consent: Logical Foundations of Constitutional Democracy* (Ann Arbor: University of Michigan, 1962).
43. For example, although some scholars might believe that federal courts are not perfectly insulated from capture, they still do not believe that federal courts are as open to capture as other government officials. See, for example, J. Jonas Anderson, "Court Capture," *Boston College Law Review* 59, no. 5 (2018): 1547.

Nature and the Firm

Jonathan H. Adler

Environmental problems are property rights problems. As Garrett Hardin noted in his seminal essay, "The Tragedy of the Commons," the creation of property rights in resources averts tragedy by encouraging sustainable use.[1] This insight, which has been made by others before and since, has been confirmed empirically. As research in various areas, from forests to fish, shows, establishing property rights in natural resources can prevent the tragedy of the commons and facilitate sustainability.[2]

Many environmental problems involve more than simple resource management, however. Many environmental problems arise out of conflicts between property owners, as when conduct engaged in by one property owner has a deleterious effect on another, or vice-versa.[3] The solution, here again, may be property rights. That is, the conflict may

J. H. Adler (✉)
Case Western Reserve University School of Law, Cleveland, OH, USA
e-mail: edolan@niskanencenter.org

J. H. Adler (ed.), *Climate Liberalism*, Palgrave Studies in Classical Liberalism,
https://doi.org/10.1007/978-3-031-21108-9_9

be solved by defining and delimiting the rights at issue, and determining which landowner has a right to what. Once this is accomplished, provided there is a legal system that will enforce the relevant rights and sufficiently low transaction costs, the system will take care of itself. Either the respective property owners will confine their activities to their own space, or they engage in voluntary transactions to reallocate the rights. Property rights will form a bulwark against pollution, and rights will be allocated to their highest and best use. Or so the theory suggests.

The problem in all but the simplest scenarios is that neither legal enforcement of property rights against pollution nor low transaction costs can be assumed. Resolving bilateral disputes between two property owners is relatively straightforward. A conflict involving multiple firms emitting pollution that affects scores of people downstream or downwind is not.[4] Mismatched or incomplete property rights further increase the difficulty of contractual or litigation-based solutions.[5]

If the transaction costs of trying to resolve pollution disputes involving many parties or other variables are prohibitive, what is the alternative? The traditional fall back is centralized regulation administered by a government agency. If it is too costly for property owners to identify the relevant sources of pollution and induce them to alter their conduct by some combination of litigation and negotiation, then a regulator will determine what is to be done. Whether the regulator adopts traditional command-and-control limitations, such as end-of-the-pipe technology controls or some sort of market-oriented emission allowance scheme, the resulting regime is often seen as a least-bad alternative to an impracticable reliance upon property rights, even if administrative regulation does little to reduce transaction costs.[6]

Are policymakers too quick to discard the property rights paradigm in favor of centralized regulation? If the available alternatives are individuated private property rights enforced through common law mechanisms, on the one hand, and centralized administrative regulation on the other, it would be hard to fault falling back on the latter where the former seems unmanageable, even if administrative regulation is far from ideal. But these may not be the only options.

Reliance upon property rights does not necessitate relying upon individuated property owners acting in an atomistic fashion nor is government intervention the only appropriate response to high transaction costs. Just as firms often utilize centralized command structures to increase production, private institutional structures may be capable of dealing with transaction costs more effectively than the government alternative. As Ronald Coase advised, "To have an efficient economic system it is necessary not only to have markets but also areas of planning within organisations of the appropriate size."[7] Likewise to have an efficient system of environmental protection, it is necessary to have areas of planning within organizations of appropriate size that otherwise exist within a decentralized and competitive system, but such planning need not be done by governmental entities. Rather such planning can occur within firms. "The Problem of Social Cost" must be supplemented by "The Nature of the Firm."

The Firm

Ronald Coase's "The Problem of Social Cost" may be the most cited—and most miscited—law review article ever published.[8] Economists, more so than legal academics, are also familiar with Coase's other seminal, and widely cited, work, "The Nature of the Firm." These articles, though superficially concerned with separate subjects, spring from a single insight: the importance of transaction costs. Indeed, Coase himself wrote that "The Problem of Social Cost and "The Nature of the Firm" "embody essentially the same point of view."[9] In both cases, his focus was on the development of institutions—in one case, the firm, in the other, the law—that reduce transaction costs and facilitate productive activity in the real world. As Coase explained:

> I showed in "The Nature of the Firm" that, in the absence of transaction costs, there is no economic basis for the existence of the firm. What I showed in "The Problem of Social Cost" was that, in the absence of transaction costs, it does not matter what the law is, since people can always negotiate without cost to acquire, subdivide, and combine rights whenever this would increase the value of production.[10]

The firm and the law are complementary institutions.

In a theoretical world without transaction costs, all factors of production would be allocated through the price mechanism and contracted for on an as-needed basis. Yet, Coase observed, this is not what we observe in the world around us. Rather we see all sorts of organizations—firms—featuring command structures that engage in "economic planning." Why would this be? Because of transaction costs; "The main reason why it is profitable to establish a firm would seem to be that there is a cost of using the price mechanism."[11] The most obvious such cost, in Coase's view, was "discovering what the relevant prices are,"[12] but that is hardly the only cost involved with carrying out a transaction in an open marketplace.[13] Entrepreneurs create firms, Coase posited, because in some cases the efficiency losses from the centralized direction of resources would be lower than the costs of transacting for the same resources. As he would write later:

> The source of the gain from having a firm is that the operation of a market costs something and that, by forming an organization and allowing the allocation of resources to be determined administratively, these costs are saved.[14]

As Oliver E. Williamson described Coase's view: "within markets the price system signals (decentralized) resource allocation needs and opportunities; but firms employ a different organizing principle—that of hierarchy—whereupon authority is used to effect resource allocations."[15]

Coase hypothesized that "firms will emerge to organize what would otherwise be market transactions whenever their costs were less than carrying out the transactions through the market."[16] The size and scope of individual firms change over time as the factors that affect the costs of organization and transacting will change over time. "Business men will be constantly experimenting, controlling more or less," creating a "moving equilibrium."[17] Just because a given firm structure is more efficient than available alternatives at time A does not mean it will continue to be more efficient than available alternatives at time B. But at any given time, it is virtually certain that there will be some amount of planning and control that will best occur within firms of various sizes and structures.

The Firm and the Environment

If, as "The Problem of Social Cost" suggested, transaction costs may inhibit the use of price signals to determine the allocation of environmental resources, particularly those utilized by large numbers—e.g., whether a water body should be utilized as a waste sink, an environmental amenity, or some combination thereof—the question is whether there are (or could be) non-governmental institutions capable of making such determinations. If, as Coase suggested, "the most important adaptation to the existence of transaction costs is the emergence of the firm,"[18] why should environmental problems not lead to the creation of environmental firms? Indeed, as Coase wrote in "The Problem of Social Cost":

> Where contracts are peculiarly difficult to draw up and an attempt to describe what the parties have agreed to do or not to do (for example, the amount and kind of smell or noise that they may make or will not make) would necessitate a lengthy and highly involved document, and where, as is probable, a long-term contract would be desirable, it would be hardly surprising if the emergence of a firm or the extension of the activities of an existing firm was not the solution adopted on many occasions to deal with the problem of harmful effects.[19]

In the study of industrial organization, the firm is recognized as a means of addressing the transaction costs inherent in market exchange. The firm is an alternative means of organizing a transaction that would otherwise occur in the marketplace.[20] Yet most discussions of environmental problems, and pollution problems in particular, proceed without any consideration of firm-like responses to the transaction costs of adopting and implementing contractual or other solutions.[21] This leads to an undervaluation of property-based institutional responses to environmental problems as compared to their regulatory alternatives.

Private ownership does not necessarily require a single, individuated owner for each resource. As Yoram Barzel advises:

> Transaction cost analysis does not end with the development of rights. The same considerations that determine when an asset will move from the public domain into private hands also suggest the organizational framework under which the asset will be exploited. Because the most valued uses of an asset usually require the cooperation of other asset owners, property rights – even when developed – are not independent of transactions involving others.[22]

From condominiums and cooperatives to partnerships and corporations, the modern economy is filled with all manner of property arrangements that enable large groups of people to join together and manage resources collectively. This is readily recognized and acknowledged in the "economic" realm, but not in environmental policy. If transaction costs are a driving concern, more attention must be paid to the full range of possible institutional and organizational structures available to coordinate human activity and resolve potential conflicts among property uses. Assuming the choices are limited to individuated property ownership on the one hand, and government regulation, on the other, does not do this.

Consider a stream subject to water pollution. If there is one upstream polluter (P) and one downstream rightsholder (R), the problem is not particularly difficult. Assuming R can identify P as the source of the pollution, R would have a legal cause of action against P for violating her rights. The legal enforceability of R's rights as against P's pollution also creates the potential for P and R to negotiate over the disposition of the rights in question. As a consequence, P must now take into account the value R places on her rights in the stream, and R must take into account the value of P's production. Whether P continues to pollute the stream will depend on whether P is willing to pay R an amount R is willing to accept as compensation for the costs imposed. Full compensation is a possibility, as is some arrangement in which P offers R a partial payment and agrees to a partial reduction of the polluting activity.

A contractual solution in which P pays R is not the only possible solution in which P continues to operate and deposit effluent in the stream. Another solution would be for P to purchase R, and thereby acquire the relevant rights.[23] At one level, this is an equivalent solution—a voluntary, wealth-maximizing transaction between two willing parties to resolve a dispute that results in the same ultimate use of the resource in question. Yet institutionally, it is different. It effectively involves the integration of the two parcels, thereby eliminating any external conflict between their prior competing uses. The trade-offs between the competing uses remain, but they are no longer the subject of legal or other dispute. The new owner is now "'incentivized' to take the interaction cost into account when deciding on the uses to which he puts his land" in pursuit of maximizing the wealth that can be achieved from the properties as a whole.[24] Yet the new owner does not have to be concerned with the transaction costs of resolving the trade-offs between potentially competing uses, as this decision can now be made within the context of a single entity—a firm.[25] Whether this solution represents an economic improvement depends upon how the organizational costs of integrating the two properties compare with the transaction costs of requiring separate owners to negotiate with each other.

The prospects of combining or integrating separate property parcels into a single unified firm suggest one possible solution to problems involving multiple users of a single resource or multiple polluters and multiple rights holders. If there are now ten polluters along the stream, and twenty rights holders, the transaction costs of a negotiated settlement are potentially quite large. If, however, all of the rights in the stream were combined into a single entity—a stream protection authority or stream basin association, perhaps—then negotiated solutions seem more possible.

A single firm representing all downstream rightsholders could facilitate the defense of rights against pollution and overcome the collective action and transaction costs that might otherwise prevent efforts to litigate against or negotiate with upstream sources of pollution. The Anglers Cooperative Association arguably played this role on behalf of riparian owners in England.[26] The ACA was formed as a membership organization to represent fishermen against polluters and others who engaged in activities that violated their riparian rights. Due to the ACA, no individual riparian owner was forced to bear the costs of investigating and initiating a legal case or pushing for a negotiated settlement. These costs were spread across all members of the organization. The membership was not geographically confined to any particular area, however, let alone a given watershed. As a consequence, the ACA was more effective at litigating on behalf of its members and obtaining compensation and injunctions, than it was facilitating contractual solutions to conflicts between riparian owners and other river users that balanced the potentially competing uses of a specific river or watershed. As a firm, it was focused on protecting fishing rights from pollution, not maximizing the value of a specific water resource.

An ACA-like entity, structured as a cooperative or condominium association, based in a given watershed or ecological area, could not only represent the interests of individual members, but represent their collective interest in the local resource as well. Much as with a homeowners association, membership could be based upon ownership of property within a watershed or along a stream, and voting shares or other interests could be apportioned equally—one parcel, one vote—or based upon the area, frontage, or waterfront of each parcel. Such a watershed authority or river basin association could not only address the collective action problem that may prevent individual rights owners from pursuing actions against polluters, but could also represent their interests as a whole. Designating a single agent to represent the interests of a group in negotiation is a demonstrated way of reducing transaction costs and facilitating contractual agreements.

This environmental firm would have the ability, and the interest, in focusing its efforts on addressing those actions which pose the greatest threat to those aspects of the watershed most important to the owners and most threatened by the activities of others. The authority could evaluate the trade-off between enforcing the rightsholders' collective interest in keeping their water free from pollution and the economic losses of restricting polluting activities—and recognize that this is not a dichotomous choice, as some non-zero amount of pollution could be worth some degree of compensation. If, for instance, most of the rightsholders are interested in protecting fish populations, they could contract with upstream polluting firms to modify their conduct in such a manner so as to reduce the threat posed to fish, and this could be done in many different ways, from limiting effluent to altering the timing or concentration of discharges, and so on.

Property owners would have an interest in the creation of an environmental firm to aggregate their interests in a collective environmental resource. Those who use the resource as an unowned or common pool, such as industrial entities whose emissions could constitute actionable pollution, may have less of an interest in forming a collective structure to facilitate negotiation and litigation. If there are many firms that are all contributing to the pollution of a given watershed, the transaction costs involved with seeking legal resolution would be high. Barring some legal device to aggregate all potential defendants (such as joint and several liability), legal actions or negotiations would have to be negotiated with each factory separately, and each factory may only be responsible for a portion of the relevant harm, reducing the value of any potential legal settlement or negotiation. This might discourage actions against any of them. While a polluting firm may aggressively seek to pull in other firms as co-defendants once it is subject to suit, prior to the initiation of any such action, a polluting firm might actually prefer the "protection" afforded by the transaction costs.

Whether or not it would be in the interest of polluting firms to join together and create a firm, basin-wide institutions could be a more efficient way of addressing certain pollution concerns. In many contexts, the pollution caused by a single firm is of much less concern than the aggregate effect of multiple polluting firms. A single factory emitting ozone precursors in the Los Angeles basin might not matter, but once there is a critical mass of such firms, their aggregated emissions will cause pollution concentrations that cause serious air-quality problems.

Where aggregate pollution levels are the concern, the costs of reducing emissions can vary from firm to firm, and even across discharge points within a single firm. This is the reason economists proposed the use of various regulatory instruments, such as "bubbles," to regulate emissions. The idea of a "bubble" is to treat firmwide emissions as if they come from a single source, and to allow the firm to determine how best to meet emission its emission reduction obligations internally. So if there are five discharge points of benzene emissions at a refinery, and the refinery is obligated to reduce its emissions by 20%, the bubble allows the refinery to determine whether it is more efficient to reduce the emissions from each source by 20%, or to concentrate the emission reductions at some subset of the sources, so long as the aggregate target is met. The creation of an environmental firm contiguous with a river basin or airshed would enable this principle to be applied across multiple facilities, but it would also facilitate greater consideration of potential alternatives.

The creation of an environmental firm could also improve the dynamic efficiency of instrument choice in pollution control. Assuming that enforcement, monitoring, and management costs do not vary across emission-control instruments, emission trading schemes and the like should be more efficient and cost-effective than their command-and-control alternatives. Allowing the collective emission reduction obligations of emitters within a given airshed to be met by reductions from those sources at which it is cheapest to reduce emissions will achieve the same reduction at less cost than imposing percentage reductions or specific control technologies across all firms. This is the conventional economic argument for emissions trading.

Yet enforcement, monitoring, and management costs are not likely to be the same for all types of emission-control instruments. Indeed, there are reasons to suspect that such costs for emission trading regimes will often be greater than the cost of technology standards.[27] For the regulator, it is relatively easy to determine whether a given firm has installed a given technology and whether it is in use. Real-time monitoring of actual emission levels, which may often be based upon internal facility operations and output, as well as the verification of emission trades may be more costly.

Whether the enforcement costs of a trading regime are greater than the benefits from the increased efficiency of utilizing such a regime will depend upon many factors, some of which are specific to time, place, and the emissions at issue. Enforcing a trading regime among a handful of facilities from which all relevant emissions are discharged from readily monitored point sources will be less costly than a trading regime focused upon non-point source emissions of a difficult-to-observe substance. Moreover, in some cases, it is easier to monitor a proxy for emissions, such as fuel use or land-use practices, than to monitor the emissions directly. Yet reliance on a proxy may reduce the cost-effectiveness of a given regime.

The choice of emission-control instrument can be seen as analogous to the "make or buy" decision faced by a firm. Sometimes it is less costly to contract for a factor of production. Sometimes it is less costly to command that it be produced within the firm. Sometimes the market alternative is less costly than the command alternative, but other times not. In a competitive economy, firms discover whether it is more or less costly to make or buy given factors of production, and the relative costs of different options change over time as a result of many factors, including technology. At one point in time, the management costs of large bureaucratic corporate structures were lower than the transaction costs of contracting for the relevant services and needs. Due to technological advances and other discoveries, this changed over time, and firm size and organization evolved accordingly.

Much like the optimal firm size and internal structure, the optimal instrument choice for pollution control will change over time, and may vary from place to place based on the pollution at issue and other factors. It is something that is best discovered in a competitive institutional environment in which the relevant decision-makers bear the costs and reap the benefits of their decisions and may simultaneously learn from observing the choices made by others.[28] Before determining which instrument is the best tool to address a particular concern at a particular time, it is necessary to identify the proper institutional context in which that decision should be made.[29]

Where Are the Environmental Firms?

If the creation of environmental firms could provide an alternative means of addressing some types of environmental problems by lowering transaction costs and avoiding some of the pathologies of administrative regulation, why have such institutions not been created? In some areas, they have been, and in others, they are discouraged or precluded by existing legal rules. In the 1970s, commenting on the development of modern environmental regulation, Allen Kneese and Charles Schulze lamented the failure to adopt regional institutions scaled to the scope of relevant ecological resources to facilitate pollution control:

> In the longer perspective—especially in light of the relationship between management of water quality on the one hand and the need to develop programs for the control of residuals from nonpoint sources on the other—the failure to build institutions that could undertake efficient region-wide management was perhaps the most profound deficiency of the entire approach.[30]

Nonetheless, there are examples of transaction-cost-reducing institutional innovations that have been used to enhance natural resource management, even if not all that often for pollution control.

The development of industrial ecology provides examples of instances in which entrepreneurs coordinated industrial developments so as to take advantage of complementary resource needs and production methods, capturing potential efficiencies and reducing potential pollution problems. In some cases, this took the form of integration, as when a firm discovered a way to make a waste product or potential source of pollution a valuable factor input for another element of production. In other cases this involved the coordination of multiple companies within a broader institutional framework.

The industrial ecosystem in Kalundborg in the Netherlands provides a useful illustration of the sort of development that can emerge:

> The Asnæs power company supplies residual steam to Statoil refinery and, in exchange, receives refinery gas that used to be flared as waste. The power plant burns the refinery gas to generate electricity and steam. It sends excess steam to a fish farm that it operates, to a district heating system serving 3,500 homes, and to the Novo Nordisk plant. Sludge from the fish farm and pharmaceutical processes becomes fertilizer for nearby farms. The power plant sends fly ash to a cement company, while gypsum produced by the power plant's desulfurization process goes to a company that produces gypsum wallboard. Finally, the Statoil refinery removes sulfur from its natural gas and sells it to Kemira, a sulfuric acid manufacturer.[31]

Interestingly enough, the Kalundborg development was not a planned result. Rather, like other examples of successful industrial ecology and inter-firm recycling, it gradually evolved over time as the result of individual contractual arrangements among the participating facilities. This illustrates the importance of allowing decentralized institutional arrangements to discover ways of economizing through coordination.[32]

The development of river basin associations, displaced by federal pollution control law in the United States but still in operation in parts of Europe, provide examples of large-scale ecological institutions capable of managing resources on behalf of a wide range of resource users.[33] The history of irrigation districts and mutual irrigation companies provides additional examples of how such institutions may evolve to coordinate the use, management, and production of common resources, or resources of sufficient scale and scope to make individual ownership impracticable.[34] The evolution of homeowners associations, cooperatives, and the like further illustrate the sort of institutional evolution and degree of variation that are possible.

Still, the question remains, if environmental firms are valuable, why have we not seen more of them? The answer (again) is transaction costs. That is, the costs of developing and managing such institutions have been greater than the benefits such institutions would provide. This explanation is somewhat tautological, however, and not particularly informative. The real question, then, is why have the transaction costs of developing such institutions been so high? One answer is that the legal environment has discouraged and in some cases prohibited the development of such institutions. Coase observed that legal institutions can make the creation of firms more or less likely by increasing or decreasing the relevant costs.[35] In the case of environmental policy within the United States, the legal environment has greatly increased the costs of institutional evolution for environmental protection. The inability to own certain environmental resources or contract for particular sorts of environmental outcomes inhibits the formation of environmental firms as there is little to be gained through the formation of such institutions.

Consider the case of water. In much of the United States, water cannot be owned until it has been withdrawn. It is not possible for an individual, let alone a corporation or an environmental firm, to own the water in place and manage it as a private resource. While there are rights associated with water, such as riparian rights, the prevailing legal system prevents the purchase and assembly of such rights into various institutional arrangements.

Under the Clean Water Act, what matters for legal purposes is whether a point source is discharging more than is provided for under its discharge permit. If discharge exceeds the permitted level, the discharge is illegal and the discharger can be subject to public enforcement and private suit, without regard for whether the discharge has any measurable ecological effect, let alone causes harm to any rightsholders. At the same time, in some jurisdictions, a valid permit may be a defense against legal claims brought by downstream riparian owners alleging harmful effects from pollution.

Due to the legal regime created by the Clean Water Act, the transaction costs for seeking a contractual or integrative resolution to a conflict between a discharger and downstream riparian owner are effectively infinite, as they cannot contract among themselves for an alternative pollution control arrangement. Even were the relevant government authority to agree to an arrangement, such as one that allows for slightly higher discharge levels in return for compensation or investments in offsetting environmental improvements elsewhere, there is no way for the regulatory authority to approve of the arrangement. Standing rules are sufficiently loose that nearly anyone can challenge the arrangement and go after the discharger for exceeding the allowable pollution level. The number of parties that would need to be part of any such agreement is, in effect, infinite, effectively precluding any contract or voluntary arrangement.

Similar problems can be seen in the context of air pollution. Local jurisdictions are required to meet federal air quality standards for local pollution levels, without regard for whether or not local pollution levels have an effect on other jurisdictions. Jurisdictions that violate these standards are forced to impose a menu of specific air pollution control measures. Although the structure of the federal Clean Air Act implies there is some flexibility in local implementation, in practice local jurisdictions are not given much choice. Listed control measures must be adopted, even if other control measures are available that could produce the same or greater emission reductions at lower cost.

At the same time that federal law prescribes the allowable pollution levels and pollution control instruments local jurisdictions must adopt, it gives local jurisdictions relatively little protection from upwind jurisdictions, particularly if a given jurisdiction seeks to reduce pollution levels below the federally proscribed standards. Interstate nuisance actions are wholly displaced for air pollution,[36] as they are for water pollution,[37] and the remedies provided through the Clean Air Act are often ineffective, and do not provide any clear remedy if an upwind jurisdiction prevents a downwind jurisdiction from meeting a lower pollution target.

In the case of waste management and disposal, the generation of certain wastes with certain characteristics in sufficient quantities automatically triggers a range of regulatory requirements that greatly increase the ability of firms to recover or recycle waste streams that can be put to productive use. Firms and property owners have limited ability to allocate environmental risks through contractual arrangements.

Environmental laws are not the only obstacles to the evolution of environmental firms, however. Antitrust law inhibits the ability to create large-scale institutions with the authority to make decisions that could influence production.[38] Antitrust also inhibited the ability of fishing cooperatives to develop rules to reduce fishery catch levels.[39]

Were many of the relevant federal environmental laws to operate as default rules that could be bargained around, rather than hard-and-fast prescriptions, there would be greater room for institutional evolution. If, for instance, rightsholders and others in a given river basin could opt out of the Clean Water Act's requirements in exchange for adopting a pollution management regime that would ensure the interests of downstream rightsholders are protected, it would be easier for such arrangements to emerge. Under the existing legal rules, however, there is little to be gained by pursuing such institutional reforms.

The Airshed authority—A Thought Experiment

Consider a hypothetical airshed governed by an environmental firm we shall call the local airshed authority (LAA). The LAA is the owner of the relevant rights in the airshed and is, in turn, owned by local property owners and governed through some form of representative system. The LAA has to determine what level of air quality is desired within the airshed, but must also ensure that pollution concentrations from within the airshed to not result in pollution downwind. In determining the level at which to set pollution levels within the airshed, LAA will take into account the cost of achieving a given pollution target, as well as the benefits of reducing pollution to particular levels, both in terms of health effects as well as in terms of the preferences of the LAA's shareholders, the local residents.

The LAA has an incentive to develop more cost-effective means of reducing pollution levels over time. Whether an emission tax, trading scheme, or some type of command-and-control regulatory standard makes sense will be function of the total costs of implementing the relevant control instruments and their expected benefits. A similar calculation will be made with the scope of control. If, for instance, most emissions come from stationary sources, and it is costly to target and control mobile source emissions, the LAA may opt to focus its efforts completely on the former. If the relative contribution of mobile source emissions increases over time, or means are developed to monitor and control such emissions at a lower cost, the LAA might opt to impose greater controls on such sources. As in other contexts, the existence of transaction costs affect the choice of instrument, and there is an incentive to reduce transaction costs where the gains from such reductions are greater than the costs of doing so.

Were the LAA to fail to control pollution levels such that pollution transport affected downwind jurisdictions, rightsholders, or firms, the LAA's existence would significantly reduce the transaction costs of resolving the dispute. Assuming the LAA lacked the right to impose pollution on those downwind, those affected by its failure to control pollution levels would not need to take action against all pollution sources within the LAA. Rather they could proceed against the LAA itself, leaving the choice of how precisely to reduce emissions for the LAA to decide. The existence of the LAA would also create the possibility of a negotiated settlement, involving compensation or some other accommodation.

The point here is not that the creation of such an authority is necessarily an optimal solution, but that it is one potential institutional arrangement worth consideration. The existence of the environmental firm creates the possibility of addressing these sorts of concerns on an airshed-to-airshed basis, rather than requiring contracting, litigation, or even regulation of each pollution source independently. Just as the creation of a firm in the economic marketplace may create efficiencies due to economies of scale in production, the creation of an appropriately scaled environmental firm may create efficiencies by reducing transaction costs and aggregating interests.

Firms and Federalism

One role for government is to solve coordination problems, such as those which could inhibit the development of environmental firms. Governments could also play a role in helping to create environmental firms, by authorizing or perhaps even mandating their creation in certain contexts so as to overcome the coordination problems inherent in seeking to assemble the relevant rights into a single entity. An airshed authority, such as that hypothesized above, would be difficult to create through voluntary contracts alone.

One might also ask whether governments, particularly sub-national governments such as individual states within a nation, could take the place of environmental firms, at least for some of the larger environmental commons. States are sometimes seen as super-firms.[40] They are large-scale institutions that perform functions through non-market means that often could not be provided through market transactions at a lower cost. This may be so, and states certainly have a role to play in helping to identify and develop the most efficient and effective environmental management institutions, but there are several reasons states may be no more than a second-best alternative to environmental firms.

For starters, one point of environmental firms is to allow for the creation of institutions that are appropriately scaled to the relevant ecological resources. State boundaries, however, are fixed, and rarely (if ever) coincide with relevant ecological boundaries. While state-level innovation and policy development may often be preferable to the federal alternative, it would be completely serendipitous were a state to be coextensive with the optimal firm size. Insofar as ideal environmental firms would be scaled to the size of relevant ecological resources, existing political boundaries will be over or under-inclusive.

Another benefit of environmental firms would be the internalization of trade-offs between competing economic and environmental considerations, both so that such trade-offs are incorporated into relevant management decisions but also that there is an incentive to find ways of overcoming the trade-offs and reconciling otherwise competing interests. While this could occur within state governments, states, possessed of plenary police powers, are omnibus entities with authority (and potential responsibility) for a near-infinite array of concerns. The likelihood that a state government will focus on ecological resources is unlikely, and this problem is magnified by the scale problems identified above. If the goal is to allow for institutions that are focused on achieving or considering environmental goals, states are no more than second best.

States are subject to interjurisdictional competition, and this can lead to the discovery and development of new approaches to environmental concerns.[41] Yet the optimal level of decentralization will often be much greater than can be provided by fifty state governments (in the case of the U.S.) that tend to operate as localized monopolies imposing a set single, uniform set of rules across the entire jurisdiction. While states compete, and have often enacted environmental reforms more rapidly than the federal government, the rate of political change is still substantially slower than can be achieved in non-governmental institutions. States are also subject to constitutional and other constraints that do not limit non-governmental entities, while also retaining the power to tax and regulate well beyond that which would be necessary to manage an environmental commons.

The greatest potential for states may be as incubators for environmental firms. If state governments, which are responsible for the front-line implementation of many federal environmental laws, were afforded greater flexibility by the federal government in goal selection, prioritization and instrument choice, it is possible some would adopt reforms that would facilitate the gradual evolution of environmental firms.[42]

At the national level, thinking of governments as firms—or thinking of government responsibility for protecting relevant commons in such terms—might also help to identify better and worse conservation strategies. An airshed authority, such as that hypothesized above, could decide to manage the airshed by charging for its use as a pollution sink, and rebating the revenues to all shareholders (local residents). This would ensure that those whose air is polluted are compensated. Shareholders could also decide whether they value cleaner air more or less than the compensation payments generated by emission fees.

Although it is quite far-fetched to imagine an atmospheric authority responsible for the global atmosphere, one might conclude that the logic of environmental firms counsels the adoption of a similar approach to greenhouse gas emissions—what is often called "tax and dividend." Under such a policy, fees (or taxes) would be imposed upon carbon emissions and the revenue would be rebated on a per capita basis. Again, such a mechanism would both provide for compensation for use of the common resource, as well as create an accountability mechanism through which it could be determined how those affected value environmental impacts and compensation on the margin.[43]

Conclusion

The existence of transaction costs associated with property rights solutions to pollution problems is the conventional justification for relying on administrative regulation to address environmental concerns. It is ironic, then, that conventional environmental regulation does little to reduce the transaction costs associated with resolving environmental conflicts and maximizing the value of contested environmental resources. To the contrary, contemporary environmental regulations tend to increase such costs.

If the goal is finding ways to reduce the transaction costs associated with pollution problems, there is a need to explore actual and potential institutional structures that could help resolve such conflicts and manage environmental resources on an appropriate scale. Much can be learned from the examination of firm-like structures that have evolved in other contexts, including for large-scale pollution problems (such as climate change) in which the creation of an environmental firm may not be possible. There is also a need for greater consideration of how existing legal institutions obstruct and may even prohibit the evolution of alternative institutional arrangements that can be used to address environmental concerns.

However inadequate or inefficient administrative regulation may be, advocates of property rights approaches to environmental problems have not sufficiently explored the cost-reducing alternatives. Coase's "The Problem of Social Cost helped frame the analysis of pollution problems." Greater consideration of "The Nature of the Firm" may help identify more optimal solutions.

Notes

1. Garrett Hardin, "The Tragedy of the Commons," *Science* 162, no. 3859 (1968): 1245 ("The tragedy of the commons... is averted by private property, or something formally like it"). Hardin was not the first to make this observation. *See, e.g.*, H. Scott Gordon, "The Economic Theory of a Common Property Resource: The Fishery," *Journal of Political Economy* 62 (April 1954): 135. The basic insight can be found at least as far back as Aristotle. *See* Airstotle, The Politics §1261.b32.
2. *See, e.g.*, Jonathan H. Adler and Nathaniel Stewart, "Learning How to Fish: Catch Shares and the Future of Fishery Conservation," *UCLA Journal of Environmental Law and Policy* 31, no. 1 (2013): 150; Christopher Costello, Steven D. Gaines, and John Lynham, "Can Catch Shares Prevent Fisheries Collapse?" *Science* 321, no. 5896 (2008): 1678; Roger

J. Sedjo, "Conflicting Signals," in *The True State of the Planet*, ed. Ronald Bailey (New York: The Free Press, 1995); Seth W. Norton, "Property Rights, the Environment, and Economic Well-Being," in *Who Owns the Environment?* eds. Peter J. Hill and Roger Meiners (Lanham: Rowman & Littlefield Publishers, 1998).
3. Absent a prior understanding of which property owner is entitled to the right at issue, the conflict is "reciprocal." *See* R. H. Coase, "The Problem of Social Cost," in *The Firm, The Market, and the Law* (Chicago: The University of Chicago Press, 1988), 96. ("We are dealing with a problem of a reciprocal nature.")
4. This argument is laid out in further detail in Jonathan H. Adler, "Is the Common Law the Free Market Solution to Pollution?" *Critical Review* 24, no. 1 (2012).
5. Monika Ehrman and Karen Bradshaw, "Complexities of Climate Governance in Multidimensional Property Regimes," this volume.
6. Indeed, in many contexts, the effect of administrative regulation is to increase transaction costs exponentially, if not infinitely. Insofar as administrative regulation precludes or prohibits voluntary exchange of property rights in environmental resources, the transaction costs are infinite. As Coase notes, "The kind of situation which economists are prone to consider as requiring corrective governmental action is, in fact, often the result of governmental action." Coase, "The Problem of Social Cost," 133.
7. R. H. Coase, "The Institutional Structure of Production," in *Essays on Economics and Economists* (Chicago: The University of Chicago Press, 1994).
8. *See* Robert C. Ellickson, "The Case for Coase and Against 'Coaseanism'," *Yale Law Journal* 99, no. 3, (1989): 611 n.2. As Ellickson notes, "Coase's name is consistently attached to propositions he has explicitly repudiated." Ibid. at 611. As Coase himself commented, "My point of view has not in general commanded assent, nor has my argument, for the most part, been understood." Coase, "The Firm, the Market, and the Law," 1.
9. Coase, "The Firm, the Market, and the Law," 1. In this remark, Coase was also referring to his 1946 essay "The Marginal Cost Controversy," *Economica* 13, no. 51 (1946).
10. Coase, "The Firm the Market and the Law," 14.
11. Coase, "The Nature of the Firm," 38.
12. Coase, "The Nature of the Firm," 38.

13. While Coase's theory of the firm is grounded in the existence of transaction costs, Coase did not use that term in his initial essay, nor did he identify many of the relevant costs that can be included under the heading of "transaction costs." *See* Oliver E. Williamson, "Introduction," in *The Nature of the Firm: Origins, Evolution, and Development*, eds. Oliver E. Williamson and Sidney G. Winter (Oxford: Oxford University Press, 1993). ("To be sure, the transaction cost particulars that Coase advanced in 1937 have withstood the test of time less well than the general argument that markets and hierarchies differ in transaction cost respects."). Yoram Barzel suggested transaction costs be defined as "the resources used to establish and maintain economic rights." Yoram Barzel, "Property Rights in the Firm," in *Property Rights: Cooperation, Conflict, and Law*, eds. Terry L. Anderson and Fred S. McChesney (Princeton: Princeton Unviersity Press, 2003).
14. Coase, "Industrial Organization: A Proposal for Research," 63.
15. Williamson, "Introduction."
16. Coase, The Firm the Market and the Law," 7.
17. Coase, "The Nature of the Firm," 55.
18. Coase, "The Firm the Market and the Law," 7.
19. Coase, "The Problem of Social Cost," 116.
20. Williamson, "Introduction."
21. There are some exceptions. For example, Allan Kneese's work in the late 1960s and early 1970s was attentive to such institutional concerns, as is some of the work of Bruce Yandle. Where economists have considered such questions of institutional design, however, it has rarely occasioned explicit consideration of firms, as such.
22. Barzel, "Property Rights in the Firm."
23. *See* Harold Demsetz, "Ownership and the Externality Problem," in *Property Rights: Cooperation, Conflict, and Law*, eds. Terry L. Anderson and Fred S. McChesney (Princeton: Princeton University Press, 2003).
24. Demsetz, "Ownership and the Externality Problem," 286.
25. Others have likewise noted that what are often termed "externalities" can occasion the merger or integration of previously competing firms. *See* Otto A. Davis and Andrew Whinston, "Externalities, Welfare, and the Theory of Games," *Journal of Political Economy* 70, no. 3 (1962). As for whether the term "externalities" adds much to the analysis, Coase (who refrained from using the word in "The Problem of Social Cost") was skeptical:

> the existence of "externalities" does not imply that there is a prima facie case for government intervention, if by this statement is meant that, when we find "externalities," there is a presumption that governmental intervention (taxation or regulation) is called for rather than the other courses of action which could be taken (including inaction, the abandonment of earlier governmental action, or the facilitating of market transactions) ....
>
> The fact that governmental intervention also has its costs makes it very likely that most "externalities" should be allowed to continue if the value of production is to be maximized .... The ubiquitous nature of "externalities" suggests to me that there is a prima facie case against intervention.

Coase, "The Firm the Market and the Law," 24, 26. *See also* James M. Buchanan and William Craig Stubblebine, "Externality," *Economica* 20 (1962) (noting that externalities are only relevant in a limited set of circumstances).

26. *See* Roger Bate, "Protecting English and Welsh Rivers: The Role of the Anglers' Conservation Association," in *The Common Law and The Environment: Rethinking The Statutory Basis for Modern Envrironmental Law*, eds. Roger E. Meiners and Andrew P. Morris (Lanham: Rowman & Littlefield Publishers, 2000).
27. *See* Robert N. Stavins, "Transaction Costs and Tradeable Permits," *Journal of Environmental Economics and Management* 29, no. 2 (1995).
28. *See* Coase, "The Nature of the Firm," 55.
29. *Cf.* Coase, "The Problem of Social Cost," 114 ("the immediate question faced by the courts is *not* what shall be done by whom *but* who has the legal right to do what").
30. *See* Allen V. Kneese and Charles T. Schulze, *Pollution, Prices and Public Policy* (Washington, D.C.: Brookings Institution, 1975), 45.
31. Pierre Desrochers, "Eco-Industrial Parks: The Case for Private Planning," PERC RS-00–1 (2000), http://www.perc.org/pdf/rs00_1.pdf.
32. *See* Pierre Desrochers, "Industrial Ecology and the Rediscovery of Inter-Firm Recycling Linkages: Historical Evidence and Policy Implications," *Industrial and Corporate Change* 11, no. 5 (2002).
33. *See* David W. Riggs and Bruce Yandle, "Environmental Quality, Biological Envelopes, and River Basin Markets for Water Quality," in *Water Marketing—The Next Generation*, eds. Terry L. Anderson and Peter J. Hill (Lanham: Rowman & Littlefield Publishers, 1997). *See also* Allen

V. Kneese and Blair T. Bower, *Managing Water Quality: Economics, Technology, Institutions* (Baltimore: John Hopkins University Press, 1968).
34. *See* Stephen N. Bretsen and Peter J. Hill, "Irrigation Institutions in the American West," *UCLA Environmental Law and Policy* 25, no. 2 (2006–2007).
35. Coase offered the example of sales taxes as a government policy that encourages firm creation, as such taxes increase the costs of market transactions relative to intra-firm transactions or commands.
36. *See* AEP v. Connecticut, 131 S.Ct. 2527 (2011).
37. *See* Milwaukee v. Illinois, 451 U.S. 304 (1981).
38. *See* Kneese, *Managing Water Quality: Economics, Technology, Institutions* (1968), 90.
39. Jonathan H. Adler, "Conservation through Collusion: Antitrust Barriers to Cooperative Fishery Management," *Washington and Lee Law Review* 61, no. 1 (2004): 49–58.
40. *See* Coase, "The Problem of Social Cost," 117. ("The government is, in a sense, a super-firm.").
41. *See, e.g.,* Paul Teske, *Regulation in the States* (Washington, D.C.: The Brookings Institution, 2004); Jonathan H. Adler, "Letting Fifty Flowers Bloom: Using Federalism to Spur Environmental Innovation," in *The Jurisdynamics of Environmental Protection: Change and the Pragmatic Voice in Environmental Law*, ed. J. Chen (Washington, D.C.: Environmental Law Institute, 2004).
42. For a discussion of one potential mechanism to facilitate state-level experimentation, see Adler, "Letting Fifty Flowers."
43. There are also reasons to believe that the transaction costs of such an approach would be significantly less than the conventional regulatory alternatives. See Jonathan H. Adler, "The Legal and Administrative Risks of Climate Regulation," *Environmental Law Reporter* 51 (2021): 10,485.

Permission, Prohibition, and Dynamism

John Thrasher

The debate around environmental policy and climate change is often framed as a debate between advocates of economic efficiency on one side and advocates of environmental protection on the other. The background assumption of both sides seems to be that environmentally sound policy is at odds with economic efficiency. Protecting the environment, in the mainstream view, requires that the government act forcefully to restrict private actors from pursuing their goals. In this sense, the market and the environment are seen as fundamentally at odds.

If correct, this tension creates serious problems for the modern, open society. The two basic institutional frameworks of the open society—democracy and capitalism—would seem to be at odds with environmental protection if this view is correct. Capitalism insofar as it favors

J. Thrasher (✉)
Smith Institute for Political Economy and Philosophy, Chapman University, Orange, CA, USA
e-mail: johnthrasher23@gmail.com

J. H. Adler (ed.), *Climate Liberalism*, Palgrave Studies in Classical Liberalism,
https://doi.org/10.1007/978-3-031-21108-9_10

the market and democracy insofar as it makes policy reliant on shifting majority power; a concern that Garrett Hardin noticed in his classic piece on the tragedy of the commons.[1] Proponents of market as well as democratic solutions to environmental problems reject this tension.

Free-market environmentalism rejects the idea that there is a fundamental tension between the market and the environment. According to Terry Anderson and Donald Leal:

> Free market environmentalism connects self-interest to resource stewardship by establishing private property rights to environmental resources. Property rights compel owners to account for the costs and benefits of their actions and facilitate market transactions that create efficiency-enhancing gains from trade.[2]

In this view, property rights are the means for internalizing the externalities of environmental damage, fusing market efficiency with environmental concern. The primary tool for the free-market environmentalist is the judiciary and the legal assignment of property rights.

Many democratic theorists also reject the tension between environmental protection and democracy. To many green political theorists, the solution to environmental problems is more democracy of a certain sort rather than less. Typically, they advocate "participatory" democracy over traditional representative democracy, but they nevertheless see the current state of environmental degradation as a failure of democratic politics, rather than as a reason to reject it.[3]

There is reason to be skeptical to both solutions to the environmental problem—what we might call the "market" and the "political" solution. Both seem to underestimate basic public choice concerns and to idealize away from the core features of the respective systems. I argue that the likely failures of both the political and market solution should lead us to reconceptualize the environmental problems that we face and embrace a different solution based on adaptation and innovation. This solution is only possible, however, if the dynamic features of an open society that drive innovation and adaptation are preserved and strengthened so that they will operate well into the future. Current approaches to environmental protection are dangerous precisely because they are likely to undermine the dynamism that will allow adaptation and innovation to occur.

The argument proceeds in several stages. First, I articulate what I take to be the key features of political and market-based approaches to environmentalism. I then look at the main problems with these approaches. Innovation, I argue, is the only plausible long-term solution to the major environmental problems that we face. Adaptation and innovation, I argue, rely on economic and cultural dynamism, which is threatened by mainstream approaches to political and, to a lesser degree market-based environmentalism.

Dynamism is at the heart of modern, open societies. Despite this, it is not well understood. My aim here is to explain what dynamism is, why it is important, and why we need to be careful to protect it when making policy. This applies, I argue, especially to environmental policy, which simultaneously undervalues and endangers the dynamism of the modern open society.

Economic Dynamism

Dynamism as I am describing it is an attempt to capture and explicate the somewhat elusive feature of modern capitalist economies that Joseph Schumpeter described as "creative destruction." According to Schumpeter, Capitalism is not a process that relies on or that leads to stable equilibria, instead, it is a dynamic process of constant change. He writes:

> The essential point to grasp is that in dealing with capitalism we are dealing with an evolutionary process...Capitalism, then, is by nature a form or method of economic change and not only never is but never can be stationary. And this evolutionary character of the capitalist process is not merely due to the fact that economic life goes on in a social and natural environment which changes and by its change alters the data of economic action; this fact is important and these changes (wars, revolutions and so on) often condition industrial change, but they are not its prime movers.[4]

Economic actors, contrary to the vision presented in mainstream microeconomics, are not primarily concerned with maximization given exogenous constraints, i.e., maximizing their bundle of goods given a budget constraint. Instead, economic agents are both responding to and creating the economic environment in which they operate. Most importantly, some of these agents are essential to driving the entire process. As Schumpeter argues:

> Nor is this evolutionary character due to a quasi-automatic increase in population and capital or to the vagaries of monetary systems of which exactly the same thing holds true. The fundamental impulse that sets and keeps the capitalist engine in motion comes from the new consumers' goods, the new methods of production or transportation, the new markets, the new forms of industrial organization that capitalist enterprise creates.[5]

The people who devise these new "methods of production," "transportation," "markets," "and forms of "industrial organization" are called *entrepreneurs*. They are, according to Schumpeter, the motive force behind the constant change inherent in capitalist economies.

> The opening up of new markets, foreign or domestic, and the organizational development from the craft shop and factory to such concerns as U.S. Steel illustrate the same process of industrial mutation—if I may use that biological term—that incessantly revolutionizes the economic structure *from within*, incessantly destroying the old one, incessantly creating a new one. This process of Creative Destruction is the essential fact about capitalism. It is what capitalism consists in and what every capitalist concern has got to live in.[6]

The core idea here is that capitalism is a process of constant change. Schumpeter uses the language of evolution arguing that mutation changes firms and methods. Mutation is a process of change that is the result of slight internal changes, in the biological case made by errors in copying that result in unexpected developments. Most mutations in nature are irrelevant or harmful and many are like this in the economy as well, but the process of mutation in capitalist economies is not strictly random. Nevertheless, there is unpredictability of the changes that will occur and how they relate to one another. We can, however, identify some core elements that make more rather than less change likely.

There are two elements within a capitalist society that drive dynamism of the sort that Schumpeter is describing. As in biological evolution, the keys are *diversity* and *selection.* Economic diversity—new products, technology, methods of organization, etc.—is driven by entrepreneurs. We call this "innovation." The selection mechanism is the market. There are considerable rewards to innovation that are selected by the market, but it is never clear what the market will want. Innovation, in this sense, is always risky. Successful innovation tends to be copied and produce new innovation. As Gerald Gaus argues, "innovation depends on previous innovations; as innovations increase, their possible modifications and new uses become ever greater."[7] Because this process of innovation builds on itself, he describes diversity in modern, capitalist societies as *autocatalytic*. As he puts it:

> [T]he more innovations there are, the greater the space of the adjacent possible to the present, which allows for even more innovations, once again expanding the space of the adjacent possible. Because innovation is autocatalytic in this way, one of the best predictors of how much innovation there will be is its present degree of diversity, for that expands the space of the adjacent possible. Diversity begets diversity.[8]

Autocatalytic diversity—which can lead to "runaway innovation"—is not only a strictly economic phenomenon. We see the same diversity in culture, science, and technology in open societies. There are several reasons for this. The first is that in open, capitalist societies, virtually all institutional settings are related to the market in some way. Scientific innovations become the basis of start-up companies, cultural change is marketed and sold. The second reason is that the runaway innovation of the market is a specific version of the mechanisms of evolution that exist elsewhere. This raises the possibility that autocatalytic diversity or creative destruction might bleed over into the politics of open societies.

Political Dynamism and the Open Society

Douglass North, John Wallis, and Barry Weingast argue that this is exactly what happens. Open societies, what they call "open-access" orders, are needed to allow for the runaway innovation and creative destruction, but innovation also encourages societies to become more open. They contrast the open society with what they call "natural states." Natural states are characterized by their need to limit access to economic and political institutions to elites for the purpose of allowing those elites to capture the rents from those institutions. This, in turn, binds the elites into a governing coalition. As they argue:

> The king or ruler only becomes powerful if he or she heads a powerful coalition. Remaining king depends on maintaining a dominant coalition that can best all rivals. Rulers are just one of many relevant actors in the dominant coalition. Focusing on the dynamic relationships of the players in the dominant coalition allows us to explicate and understand the logic of the social order and the conditions underlying all social organizations in a natural state.[9]

The logic of the natural state is that it must be closed, it can't allow for the autocatalytic diversity that causes runaway innovation and creative destruction. Successful innovation creates excess returns, what economists call "rents." The natural state remains stable by controlling access to the economy and political system to distribute rents to its preferred elite coalition. Preventing access to organizations, markets, and trade, however, also dampens the autocatalytic diversity that leads to runaway innovation. As North, Wallis, and Weingast point out, "Open entry and access to sophisticated economic organizations are prerequisites for creative destruction and a dynamic economy."[10] They go on to explain that "The natural state cannot support creative destruction because the creation of new economic organizations directly threatens existing economic organizations and their patterns of rents."[11] The implication is that a dynamic economy requires an open society.

The forces unleashed by autocatalytic diversity will tend to undermine closed societies. Those who benefit from the rents developed through economic innovation will form new political coalitions and try to preserve their access to the market and to economic organizations. The mature version of the open-access political order is liberal democracy. Democratic because it is based on the open competition of various political interests and liberal because it includes the protection of political losers and minorities.

We can now see how liberal democracy mirrors the capitalist economy in an open society. Both are based on open access, and both rely on processes of autocatalytic diversity and selection through creative destruction. In democratic politics, selection occurs in elections as well as internally in parties and within bureaucracies. Political entrepreneurs, like their economic counterparts, must anticipate what the market wants and innovate to outflank their competitors.

We might think that one relevant difference between the two is that political competition is zero-sum, while economic competition is not. William Riker, reflecting this view, argued that politics, not economics was the "truly dismal science," since there was no guarantee that political competition would lead to mutual advantage.[12] There is surely a sense in which Riker is right about this; politics is about winning and losing. There is also a sense in which this view is too narrow. According to James Buchanan, democratic politics can be thought of as a kind of exchange.[13] Democratic politics involves members of a society giving up their power to unilaterally exploit and expropriate one another to gain the benefits that come from peace. These benefits come in several forms. One, which North, Wallis, and Weingast emphasize, is that democratic, open-access orders, by enlarging the ruling coalition beyond a small group of elites, are able to make credible commitments that closed societies cannot. Specifically, they can credibly commit to not expropriating the value of economic actors and institutions for political reasons. This not only encourages economic activity but also allows the government to finance large projects since it is a more trustworthy borrower.

Second, state capacity tends to be higher in democratic, open-access orders than in closed, natural states. In measures of state capacity, Jonathan Hansen and Rachel Sigman find that state capacity measures correlate highly with measures of government effectiveness, rule of law, regulatory quality, impartial public administration, and other measures that we would identify with stable, good government.[14]

All of this is to say that although political and economic dynamisms are not the same, they are clearly related. The core relation between them seems to be the autocatalytic diversity and creative destruction that is found both in capitalism and in liberal democracy. It should be no surprise then that wherever we find capitalism we also tend to find liberal democracy and vice versa. This fact may be puzzling to democratic theorists who often see capitalism as opposed to democracy[15] as well as to defenders of capitalism who are skeptical of democracy.[16] There is no mystery, however, when we see that dynamism is at the heart of both institutional forms.

Dynamism as a Social Value

The argument so far has made the case that dynamism is at the heart of the open society. We have already seen that dynamism is valuable insofar as it is tied up with innovation. There are several ways to respond to value, however. How we think about the value of dynamism will affect the policies we choose to pursue. Following Philip Pettit, we can distinguish between *promoting* and *honoring* a value.[17] This distinction tracks the emphasis of consequentialist and deontological theories of value generally. According to Pettit, the consequentialist believes that the proper response to any value is to promote it, the deontologist denies this.[18] Crucially, the deontologist need not deny that some *values* are properly promoted, the key distinction is that the deontologist denies that all values should be responded to in this way.[19]

We can see this distinction in relation to the value of dynamism. It is not immediately obvious whether we should be consequentialists about dynamism—whether we should promote it—or whether we should instead honor it in some way that is distinct from promoting it. If we think of dynamism as a value to be promoted, our goal might be to look for ways to increase dynamism. That is, ways to stimulate diversity, innovation, and creative destruction. We can value dynamism directly as a value in itself, or instrumentally, as something that facilitates or encourages other social values. In some cases, valuing something directly is also a way of valuing it instrumentally. For instance, we may value the dignity of individuals and their basic human rights in themselves but doing so may also be the best way to secure the benefits that come from a system of basic human rights and dignity. The same might be true of dynamism as well. Since dynamism is built on innovation, its fruits will be impossible to predict. Some of them may even be bad. Nevertheless, we can think about the general benefits of an open society built on dynamism versus one that is not and conclude that dynamism is, on net, valuable because of the benefits it is likely to bring, even though we may not know precisely what those are. Valuing dynamism as something to be promoted is compatible with valuing it directly if the source of the value is the good we expect dynamism to bring. Valuing it is, in this sense, an indirect consequentialist strategy rather than an alternative to consequentialism.

A true alternative to a consequentialist approach to dynamism would mean valuing it even if doing so decreased dynamism. It is hard to know exactly what that would look like in this context. The best analogy may be to free speech or toleration, which are both drivers of dynamism. Valuing freedom of speech or tolerance may not mean always promoting more speech or more tolerance. It might also mean respecting the right of the intolerant to make their case for intolerance or a restriction in speech. Both the Communist Party and various neo-Nazi groups call for the overthrow of the United States government. If they were successful, they would both limit freedom of speech and the toleration of dissenting views, though obviously in different ways. The value of freedom of speech and toleration, at least as it is understood in the United States, prevents the law from silencing either, even though they would both eliminate that very same standard of toleration and freedom if they were in power. In this sense, tolerating the intolerant is a way of treating the values of freedom of speech and toleration deontologically.

In the political domain, this might mean tolerating and defending groups that are opposed to various aspects of dynamism, for instance, traditionalist conservatives and socialists. In the economic realm, it might mean defending innovations that in the long or short-run lead to less dynamism, for instance, firm consolidation and various protections for intellectual property.

Given what we have seen about dynamism and its link to capitalism and democracy, the best way to respond to the value of dynamism is likely a two-level strategy that incorporates both a consequentialist and non-consequentialist approach. At the level of individual decisions within an economic or political context, we should not attempt to maximize dynamism. Instead, in an open and dynamic society, we should respect the rules and norms of the system as a whole. When we are evaluating the system, however, we should think about the rules and norms in terms of which are likely to lead to more rather than less dynamism. We can think of this approach as like the one that Rawls advocated in "Two Concepts of Rules."[20] There he distinguishes between what he calls a "practice" and a "summary" conception of rules. Practice rules are the rules that establish a practice, defining roles and permissible/impermissible actions within the system of conventions. Summary rules are rules that summarize or embody past learning. These are what we sometimes call "rules of thumb" or heuristics. In the context of dynamism, we should construct and change institutional rules to encourage dynamism. Individual actors need not encourage dynamism directly or aim at dynamism directly. This raises the question of what type of rules tend to encourage dynamism. That is the subject of the next section.

Prohibitions and Permissions

To simplify considerably, we can think of two general categories of rules; rules that let people know what actions are permissible and what actions are prohibited. Call these *permission* and *prohibition* rules respectively. We can think of a society as having different default approaches to institutional rules in the sense that when we are evaluating policy or looking at problems we want to solve, there is a presumption in favor of one type of rule or another. In real societies, this will vary based on specific contexts and precedents, but it is worth thinking about them at the level of a whole society to see the general features of the different types of rules.

These rules can be related to fundamental rights, such as the default rule that one can practice one's religion in America without seeking permission, or they can result from some lower-level rule. Default rules may be the result of laws or regulations, or they may inform laws or regulations. As an example, think of the default rule against discrimination in business. This rule is implemented in several specific pieces of legislation, including Title VII of the Civil Rights Act of 1964, The Equal Pay Act of 1963, and the Pregnancy Discrimination Act, among others. The general rule is that businesses are prohibited from discriminating on the basis of race, color, religion, sex (including pregnancy, sexual orientation, or gender identity), national origin, disability, age, or genetic information and medical history. The default rule can be simplified as discrimination is prohibited unless it is essential for the job.

Default rules, to be effective must meet two basic criteria. First, they must meet what we can think of as a "publicity" condition. This means that they need to be clear and articulable, and it should be common knowledge that the rule is in effect. This condition is important because, for the rule to structure policy and action, it is important for both the citizens and the special agents of the government to be able to understand the rule and know when it is being applied. One default rule will be preferred to another similar rule if it is able to meet this condition more easily. In general, this will mean that simpler and clearer rules are preferred over complex and vague rules.

Second, default rules should be decisive. This means that they should clearly rule in or rule out the available options. Gerald Gaus and Shaun Nichols define this condition as "for any action φ, [the rule] allows, requires, or prohibits φ (with the 'or' being exclusive)."[21] This can be seen as an extension of the publicity condition regarding the output of the rule. If the rule is not decisive, it won't be clear what the rule requires, allows, or prohibits. Of course, many rules—perhaps all rules—are not decisive in a logical sense in new or boundary cases. This does not necessarily undermine the decisiveness condition so long as the rule reliably directs the action of those to whom it is meant to apply. In new cases or contexts or in boundary cases, the rule may require interpretation, but the rule itself should not require interpretation or the arbitrary will of some official for it to direct action.

This condition has several important properties that make it attractive. For instance, no inherently inconsistent rule can meet this condition. If a rule directs one to both φ and not-φ, the rule is clearly not decisive. The more types of cases and contexts the rule applies to, the more general the rule will be. More general rules will tend to be better than less general rules on grounds of both publicity and decisiveness. This means their properties and attractiveness of them.

In the environmental context, the Precautionary Principle (PP) is a popular default rule. Although it has many different specifications, in all its forms is a general rule of prohibition on the action until certainty or widespread consensus is reached that a particular action or course of action is safe. One example of a rule that used this principle is found in Article 15 of the 1992 Rio Declaration:

> In order to protect the environment, the precautionary approach shall be widely applied by States according to their capabilities. Where there are threats of serious or irreversible damage, lack of full scientific certainty shall not be used as a reason for postponing cost-effective measures to prevent environmental degradation.

Put in the negative, as the Rio Declaration does, the rule states that scientific disagreement or uncertainty does not count as a reason to ignore the possible negative outcomes of an action.

One way to think of this principle is as a broadening of all of the outcomes that should be included in the evaluation of decision-making in contexts of uncertainty. As Katie Steele puts it, "Given any particular course of action, the PP calls for acknowledgment of all potential outcomes, even those that are scientifically uncertain."[22] Understood this way, the PP is a principle for identifying outcomes that should go into a traditional expected utility or cost–benefit analysis calculus. This is somewhat misleading, however. The PP is only deployed in contexts where there are potentially very negative outcomes. If we disregard the expected likelihood of those outcomes or, in cases of uncertainty, are unsure what those likelihoods are, the PP allows the potential costs of a given course of action to swamp its potential benefits in many cases. For instance, the PP has been used to block the adoption of various technologies, which may have had considerable positive impacts. Some of these cases included DDT and GM Foods:

> Among the technologies against which the precautionary principle has been invoked are DDT, a pesticide that has had spectacular success in reducing one of nature's dread diseases—malaria—worldwide, but that also has been associated with declines in the population of various avian species, such as the bald eagle and the peregrine falcon; fossil fuel combustion, on which much of the world's current prosperity and human well-being is based but which could contribute to global warming; and GM crops, which promise reductions in global hunger and malnutrition and a less chemical-dependant agriculture but which have also raised the specter of "frankenfoods."[23]

According to Indur Goklany, the PP is apt to ignore the opportunity cost of prohibition by privileging the immediately seen dangers to unseen and long-term harms and benefits. As Goklany argues:

> This one-sided application of the precautionary principle can be attributed to the fact that the principle itself provides no guidance on its application in situations where an action (such as a ban on GM crops) could simultaneously lead to uncertain benefits and uncertain harms.[24]

This means that the PP is non-decisive when we factor in all the harms and benefits, but also that it is unsuited for the very cases where it is meant to apply, namely cases of uncertainty in policymaking. We can go further than Goklany on this point, though. The PP, used as a default rule, does not pick out a particular action or policy. In this way, it is unlike a Maximin or Minimax regret decision rule. Instead, it acts as a prohibition on all actions that meet its criteria. If any actions are left that are not prohibited by the PP, an additional specific decision rule will be needed to select one. Nevertheless, the PP clearly meets the publicity condition—in that it is easy to understand and articulable.

Although it meets the first basic desiderata of a default rule, the PP has some clear drawbacks. The first is that any policy that has a potentially catastrophic result, regardless of how unlikely, will be prohibited by this rule. The problem here is that only causally or logically impossible events have zero probability, so there are a huge number of possible negative outcomes, even catastrophic ones, that might result from many actions.

The PP also doesn't rule out possible, though implausible results of action. Perhaps in some cases, this makes sense, but it is very restrictive as a general default rule. Maybe it makes sense to assume that nuclear power plants should be protected against both an earthquake and a tsunami. In many parts of the world, these are improbable events, say in the middle of Nevada, but not in Japan. Does it make sense, though, to make sure that nuclear plants are safe in the middle of a warzone, as is the case now with regard to two reactors in Ukraine? Perhaps. The cores of new nuclear plants are hardened to withstand direct hits by airliners, for instance. This was a possible, but not a plausible threat until September 11, 2001. Still, there are innumerable possible threats that are *possible* with any technology. It is possible that the invention of the computer in the 1830s by Charles Babbage could eventually lead to the creation of humanity destroying AI. Many people who study this issue even think it is plausible that this will happen within the next 50 years.[25] Is this a reason to ban developments in computing? Perhaps, but it is obvious that such a ban would foreclose huge benefits. Even the small chance—really any chance—of AI developing and destroying humanity is enough to limit the technology by the PP.

The PP also presumably applies to solutions as well as to technologies, which could potentially limit the possibility of reducing damage or already existing harms. As Neil Manson notes, nuclear power may be an important part of the solution to transition our economy from dirtier fossil fuels, but the PP rules it out on the basis of the possible harms that might occur from a series of catastrophic meltdowns and leakage of waste.[26] Of course, catastrophic climate change is also a problem, though, and if transitioning to nuclear power is the best way to prevent catastrophic climate change, it might make sense to live with the risks of nuclear power, rather than accept the risks of climate change. The PP rules out this possibility though and others like it. This example highlights a problem with general prohibition default rules, namely that they tend to exclude too many options as impermissible. It will exclude both very risky and only somewhat risky actions from the set of permissible acts if both have the possibility of leading to negative outcomes.

Permissions, Innovation, and the Environment

If the PP seems too restrictive, the opposite type of rule, such as a default rule of permission would likely not be restrictive enough. Let's consider such a rule. Call it a rule of permissionless innovation (PI). Under this default rule, any action φ is permissible so long as there is not some other established rule that prohibits it. This rule meets the desiderata stated above for default rules. It meets the publicity condition since it is clear and easy to understand how it applies. It is decisive since every action is permissible except those already prohibited. The danger with this type of rule is the flipside of the PP, innovative policies might carry risks that make them too costly to implement. To evaluate this possibility, we need to think about how likely such a result would be in an environmental context.

A PI rule is more conducive to dynamism and innovation, but we can expect some of this innovation to lead to environmental harm. It is impossible to know when we are dealing at this level of generality, but if the past is an indication, we can assume that harm is likely. Recall, though, that autocatalytic diversity and runaway innovation are likely when dynamism is encouraged. If so, we can expect solutions to arise from a system of PI as well as harm. It can't be known in advance what these harms and solutions will be, but a system based on PI—encouraging dynamism—has a selection mechanism that should tend toward eliminating unwanted externalities and harms. The more dynamic a system, the more responsive it is likely to be and the more likely we are to see solutions develop quickly to problems.

A general PI rule in most environmental contexts has several benefits. Recall that the cost of prohibition rules is that we cannot really know all the good options that were prevented by it. The harms that arise from actions under a PI rule are visible, which allows them to be more successfully mitigated and reduced. Also important is the effect on dynamism in a capitalist economy. Dynamism requires an institutional structure that encourages and allows for innovation; without it, we risk stagnation. According to Arthur Diamond:

> We are neither preordained to be blessed with bounty nor to be stuck with stagnation. It can go either way. If we are open to innovative dynamism and allow entrepreneurs to innovate, we will have bounty. If we are closed to innovative dynamism and bind entrepreneurs, we will have stagnation.[27]

PI is a rule that allows for a maximal opportunity to innovate. This should be ultimately good for the general welfare in terms of economic benefits, as well as good for developing institutional and technological solutions to environmental problems like climate change. Indeed, according to Diamond, a dynamic capitalist economy is essential to solving the problems that arise due to climate change.

> The economy in a system of innovative dynamism flourishes by allowing innovative entrepreneurs to nimbly adapt to and make use of unexpected changes in their economic environment. The flourishing economy can be sustained in the face of global warming by similarly allowing innovative entrepreneurs to adapt to and make use of changes in their natural environment.[28]

Climate change is coming whatever we do. We don't know the magnitude of the effects that we should expect, or where they will occur. We don't even know how to weigh the harms and benefits across time.[29] Many economists recommend a social discount rate of 2%.[30] Even so, there is a dispute on this point and even if there were not, we are talking about the aggregated cost of climate change across the entire society, surely something hard to be very confident about. The same is likely true with regard to setting a Social Cost to Carbon.[31] Aside from disputes about what the cost should be set at, the fact that we are arguing about the true costs indicates that we are not relying on the usual mechanism for setting prices, namely markets. Without a bargaining or market mechanism for generating prices, familiar Coasean and Hayekian problems for determining costs arise.

Given the irreducible uncertainty that we face regarding climate change and environmental risk more generally, there is no straightforward way to evaluate costs and benefits. Under conditions of risk, however, standard expected utility maximization is usually thought to be the rational individual decision rule, with some version of cost–benefit analysis being the typical policy analogue. Under uncertainty, there is no clear consensus as to what the proper rational decision rule should be.

More precisely, decisions under uncertainty obtain when we must choose from a set of possible acts $A_{1\ldots m}$ given some unknown state of the world $S_{1\ldots n}$, where the outcome is some pair of actions and states (A_i, S_j) with all possible outcomes define in $A \times S$.[32] Let's say I want to go on a vacation and my options are to go to Hawaii, Yosemite, or Colorado. Imagine further that I am making these plans during the 2020–2021 Covid-19 Pandemic and there is a chance that Yosemite or Hawaii will restrict access to visitors to slow the spread of disease. Since I am making my plans several months in advance, I don't know whether the vacation spot will be closed or open. If I choose Hawaii, I need to buy a ticket in advance and if I can't go, I will lose some money on the ticket. Since I live in California, I can drive to Yosemite so if it closes, I won't be able to go on vacation, but I also won't lose any money. Colorado can't close, so I will be able to go there regardless. I would rather go to Hawaii than Yosemite and Yosemite rather than Colorado, all things considered (Hawaii $\succ$ Yosemite $\succ$ Colorado). Given that I don't know whether the locations will be closed or open, what should I choose? (Table 1).

If I had an idea of how likely one of the locations is likely to be closed, I could just discount the outcomes for each act by their likelihood of happening and choose based on the highest expected utility. But, in case of uncertainty, we don't have any information (or any good information) about how likely each state of affairs is. Since I can't apply traditional expected utility analysis, I need another decision rule.

A popular rule is a *maximin*, which states that one should maximize the minimum outcome one can get from their choice. One assumes that

Table 1 Vacation decision matrix

	Closed to visitors	
	Yes	No
Hawaii	Lose Money and No Vacation	Great Vacation
Yosemite	No Vacation	Good Vacation
Colorado	Good Vacation	Good Vacation

the worst state of affairs obtains and then chooses the best option for that case. This ensures that even in the worst possibility, one does as well as is possible. In this case, it would mean choosing Colorado because since Colorado can't close, I am assured of getting a good vacation regardless of whether the other locations choose to close. But notice that this also means that I choose my least preferred, but the safest outcome. John Rawls famously argued that the maximin rule should be used by representative choosers behind the veil of ignorance when choosing principles of social justice. Rawls gives several reasons for this, but the main concern is to avoid choosing a set of principles that avoid the dangers that might come from highly inegalitarian or repressive principles of justice.[33] John Harsanyi famously disagreed with Rawls on this point.[34] Their disagreement didn't really hinge on technical questions in decision theory, instead, it focused on the question about the way we think the moral case in question should be represented.[35]

Notice that my case from above doesn't have complete ignorance about the ranking of outcomes given possible states. That information is not enough to make a full probability distribution that would be very reliable, but I am still able to ordinally rank the outcomes. There may be cases where I rank several outcomes as disastrous, but where one disastrous outcome is slightly better than another. In those cases, maximin may pick out an option that could lead to a disastrous outcome over one that might not. This leads to a modification of the maximin rule for when we are dealing with several very similar disastrous outcomes, a rule that Gregory Kavka called the "disaster avoidance principle."[36] This principle says to choose "the course of action that maximizes your chance of avoiding all disastrous outcome."[37] A similar rule known as *minimax regret* was proposed by Leonard Savage.[38] This rule was later extended by Graham Loomes and Robert Sugden.[39] The idea here is to minimize the difference between what one would have gotten given the best state of affairs and the actual state of affairs for a given action. To do this, one needs to be able to not only evaluate outcomes ordinally but also cardinally since we need to know the difference between actual and maximum outcomes. Since we only need ratios, we only need a utility function that preserves the relative differences between outcomes. As such, we could a Von-Neumann-Morgenstern utility function generated from subjective utilities over gambles. The rule, as Savage points out is similar to the solution to Von-Neumann and Morgenstern's minimax rule in zero-sum games where the unique equilibrium solution is to choose the strategy that minimizes your opponent's maximum payoff.[40]

All these rules aim at reducing one's exposure to downside risk rather than in maximizing expected gain. This is a feature that they share with general prohibition default rules like the PP. There are, however, different interpretations of this principle and its relation to decision rules. Minimizing downside risk in a default rule is not the same as eliminating the possibility of bad outcomes, though. We may misidentify the most serious risk or focus on problems that, though serious, are not urgent. We saw an example of the importance of dynamic innovation that a general default rule of PI can have during the recent pandemic. Social distancing was only possible because of the widespread access to technology that allowed many people to work from home. Many people were able to avoid shopping in crowded stores because of delivery services like Amazon. Most importantly, innovative research into vaccine technology and production made it possible to develop and distribute a vaccine quickly. Meanwhile, the delays that occurred were driven by PP style prohibition default rules instituted by the FDA and the CDC.[41]

Capitalism driven by creative destruction and entrepreneurial discovery ensures uncertainty since it is unforeseen solutions to existing problems that create the opportunities for gains that entrepreneurs seek. As such, it is impossible to predict the workings of a dynamic capitalist economy or how technological and institutional systems will develop within it. Capitalism is characterized by both path-dependence and increasing returns dynamics that make it impossible to predict the effects of many actions or innovations.[42] Despite what many people think, the development of the capitalism has largely been a good thing for the environment and there is reason to think that this will continue, but that does not mean that we do not face considerable risk as well.[43] It is essential that when we confront the dangers and uncertainty of our impact on the environment that we seek to preserve both the natural environment that makes our life on earth possible, but also the institutional environment of dynamic capitalism that is essential to our well-being and to finding solutions to future problems. Given the clear benefits of dynamism, we should favor PI style permission rules over PP style prohibition rules all things being equal. Of course, all things are not equal, and there may be instances where a strong case can be made for PP style rules. Even so, in those instances, we are likely to have a better sense of the outcomes and their likelihoods, which means we should engage in traditional cost–benefit analysis rather than favor a PP.

Conclusion

We act in a world of considerable uncertainty. We can rarely be certain about all the effects our actions will have and we are also often faced with situations where we don't know what we don't know. In the face of uncertainty, decision theorists have developed several possible decision rules, with no consensus on which are the most rational. In an institutional context, we need default rules that tell us which actions are permissible, impermissible, or prohibited. Two default rules of interest are a general prohibition rule in the form of the Precautionary Principle and a general permission rule in the form of a Permissionless Innovation. I have argued that given the importance of dynamism in our open society, we should tend to prefer PI style rules over PP rules.

Notes

1. Garrett Hardin, "The Tragedy of the Commons," *Science* 162, no. 3859 (December 13, 1968): 1243–1248, https://doi.org/10.1126/science.162.3859.1243.
2. Terry L. Anderson and Donald R. Leal, *Free Market Environmentalism for the Next Generation* (New York, NY: AIAA, 2015), 3.
3. Robert E. Goodin, *Green Political Theory* (John Wiley & Sons, 2013); Hélène Landemore, *Open Democracy: Reinventing Popular Rule for the Twenty-First Century* (Princeton: Princeton University Press, 2020).
4. Joseph Schumpeter, *Capitalism, Socialism and Democracy*, 5th Revised ed. (London: Routledge, 1942), 82.
5. Ibid., 82–83.
6. Ibid., *Capitalism*, 82.

7. Gerald Gaus, *The Open Society and Its Complexities* (Oxford: Oxford University Press, 2021), 114.
8. Ibid., 114.
9. Douglass C. North, John Joseph Wallis, and Barry R. Weingast, *Violence and Social Orders: A Conceptual Framework for Interpreting Recorded Human History* (Cambridge University Press, 2009), 31.
10. North, Wallis, and Weingast, *Violence*, 23.
11. Ibid., 116.
12. William H. Riker, *Liberalism against Populism: A Confrontation between the Theory of Democracy and the Theory of Social Choice* (Prospect Heights, IL: Waveland Press, 1988), 206.
13. James M. Buchanan, *The Reason of Rules, The Collected Works of James M. Buchanan* (Indianapolis: Liberty Fund Inc., 1985), 116.
14. Jonathan K. Hanson and Rachel Sigman, "Leviathan's Latent Dimensions: Measuring State Capacity for Comparative Political Research," *The Journal of Politics* 83, no. 4 (October 2021): 1503–1505, https://doi.org/10.1086/715066.
15. Thomas Christiano, "The Uneasy Relationship between Democracy and Capital," *Social Philosophy and Policy* 27, no. 1 (January 2010): 195–217, https://doi.org/10.1017/S0265052509990082.
16. Bryan Caplan, *The Myth of the Rational Voter: Why Democracies Choose Bad Policies* (Princeton University Press, 2011); Jason Brennan, *Against Democracy* (Princeton: Princeton University Press, 2016).
17. Philip Petttit, "Consequentialism and Respect for Persons," *Ethics* 100, no. 1 (October 1989): 116–126.
18. Ibid., 119.
19. Gerald Gaus, "What is Deontology? Part One: Orthodox Views," *The Journal of Value Inquiry* 35, no. 1 (2001): 27–42; Gerald Gaus, "What is Deontology? Part Two: Reasons to Act," *The Journal of Value Inquiry* 35, no. 2 (2001): 179–193.
20. John Rawls, "Two Concepts of Rules," *The Philosophical Review* 64, no. 1 (January 1955): 3–32.
21. Gerald Gaus and Shaun Nichols, "Moral Learning in the Open Society: The Theory and Practice of Natural Liberty," *Social Philosophy and Policy* 34, no. 1 (July 2017): 87, https://doi.org/10.1017/S0265052517000048.
22. Katie Steele, "The Precautionary Principle: A New Approach to Public Decision-Making?," *Law, Probability and Risk* 5, no. 1 (March 1, 2006): 21, https://doi.org/10.1093/lpr/mgl010.

23. Indur M. Goklany, *The Improving State of the World: Why We're Living Longer, Healthier, More Comfortable Lives on a Cleaner Planet*, Annotated edition (Washington, DC: Cato Institute, 2007), 238.
24. Ibid., 238–239.
25. Nick Bostrom, *Superintelligence: Paths, Dangers, Strategies*, Reprint edition (Oxford, United Kingdom; New York, NY: Oxford University Press UK, 2016).
26. Neil A. Manson, "Formulating the Precautionary Principle," *Environmental Ethics* 24, no. 3 (2002): 273, https://doi.org/10.5840/enviroethics200224315.
27. Arthur Diamond, *Openness to Creative Destruction: Sustaining Innovative Dynamism* (Oxford University Press, 2019), 4.
28. Ibid., 122.
29. William D. Nordhaus, "Discounting in Economics and Climate Change: An Editorial Comment," *Climatic Change* 37, no. 2 (October 1997): 315–328, http://dx.doi.org.ezproxy.lib.monash.edu.au/10.1023/A:1005347001731.
30. Moritz A. Drupp et al., "Discounting Disentangled," *American Economic Journal: Economic Policy* 10, no. 4 (November 2018): 109–134, https://doi.org/10.1257/pol.20160240.
31. Laurie T. Johnson and Chris Hope, "The Social Cost of Carbon in U.S. Regulatory Impact Analyses: An Introduction and Critique," *Journal of Environmental Studies and Sciences* 2, no. 3 (September 2012): 205–221, https://doi.org/10.1007/s13412-012-0087-7.
32. R. Duncan Luce and Howard Raiffa, *Games and Decisions: Introduction and Critical Survey* (New York: John Wiley & Sons, 1957), 276–277.
33. John Rawls, *A Theory of Justice*, Revised (Belknap Press, 1971), 134.
34. John C. Harsanyi, "Can the Maximin Principle Serve as a Basis for Morality? A Critique of John Rawls's Theory," *The American Political Science Review* 69, no. 2 (June 1, 1975): 594–606, https://doi.org/10.2307/1959090.
35. Gerald Gaus and John Thrasher, "Rational Choice in the Original Position: The (Many) Models of Rawls and Harsanyi," in *The Cambridge Companion to the Original Position*, ed. Timothy Hinton (Cambridge: Cambridge University Press, 2015), 39–58; Michael Moehler, "The Rawls–Harsanyi Dispute: A Moral Point of View," *Pacific Philosophical Quarterly* 99, no. 1 (2018): 82–99, https://doi.org/10.1111/papq.12140.

36. Gregory S. Kavka,"Deterrence, Utility, and Rational Choice," *Theory and Decision* 12, no. 1 (March 1980): 41–60, https://doi.org/10.1007/BF00154657.
37. Gregory S. Kavka, *Hobbesian Moral and Political Theory* (Princeton University Press, 1986), 143.
38. L.J. Savage, "The Theory of Statistical Decision," *Journal of the American Statistical Association* 46, no. 253 (March 1951): 55–67, https://doi.org/10.1080/01621459.1951.10500768.
39. Graham Loomes and Robert Sugden, "Regret Theory: An Alternative Theory of Rational Choice Under Uncertainty," *The Economic Journal* 92, no. 368 (1982): 805–824, https://doi.org/10.2307/2232669.
40. Savage, 63; Jon von Neumann and Oskar Morgenstern, *Theory of Games and Economic Behavior* (Princeton: Princeton University Press, 1944), 641.
41. Scott Gottlieb, *Uncontrolled Spread: Why COVID-19 Crushed Us and How We Can Defeat the Next Pandemic* (Harper, 2021).
42. Paul Pierson, "Increasing Returns, Path Dependence, and the Study of Politics," *American Political Science Review* 94, no. 2 (June 2000): 251–267, https://doi.org/10.2307/2586011.
43. See Daniel Halliday and John Thrasher, *The Ethics of Capitalism* (Oxford: Oxford University Press, 2020).

Market Solutions to Large Number Environmental Problem-Induced Changes in Risk Distributions

Andrew P. Morriss

Introduction

We usually think of insurance as a way to shift losses between parties to diversify risks.[1,2] For the environment, this is often "pollution liability insurance" which "helps protect businesses from unexpected pollution exposures that may not be covered by standard casualty and property policies."[3] This goes beyond helping businesses at risk of causing environmental harm shift financial risks.[4] More advanced risk transfer mechanisms (ARTMs) including captive insurance, insurance-linked securities, and insurance derivatives can contribute to addressing the problems by encouraging dynamic, adaptive responses to risk and uncertainty by generating and applying knowledge. This chapter argues that by developing methods of risk management, which require investment in developing information resources on climate risks, we will produce

A. P. Morriss (✉)
Texas A&M University, College Station, TX, USA
e-mail: amorriss@tamu.edu

J. H. Adler (ed.), *Climate Liberalism*, Palgrave Studies in Classical Liberalism,
https://doi.org/10.1007/978-3-031-21108-9_11

a better-informed approach to managing them. Using this information, entrepreneurs across firms will innovate to address them. ARTMs are an important means of producing a dynamic response to climate risks. Markets not only provide an important mechanism for individuals and firms to adapt to climate change but they also produce the information needed for a better understanding of climate risks.

I argue that ARTMs offer a market-based mechanism that can help solve the class of environmental problems involving which I clumsily label "large number environmental problem-induced changes in risk distributions," such as the risks associated with climate change. These problems are different from the environmental problems that expose individuals or limited areas to a risk whose distribution is (at least partially) known. They differ in kind from problems, such localized harm caused by a chemical plant, where the harm—while significant—is to those in the vicinity of the plant. For example, a chemical release at the BP Texas City refinery in 2007 injured workers at the plant.[5] Ten workers were initially awarded $10 million in punitive damages each (later reduced to less than $500,000 total) as well as compensatory damages of $5000 to $240,000. OSHA fined BP $87 million for failing to correct safety hazards after a 2005 explosion that killed 15 people and injured 150.[6] Further lawsuits were filed against BP over later incidents at the same plant.[7]

Problems like BP's can be addressed through the regulatory system and private tort lawsuits. While there are issues with the specifics of the regulation of the plant and the legal system's handling of the tort claims, such problems are what regulatory agencies and tort suits are meant to address.[8] The fact issues can be difficult—Was a toxic substance released? Whose fault was it? Did it harm those exposed and by how much?—but these are not categorically different from the problems regulatory agencies and courts routinely handle.

Similarly, the Flint, Michigan residents exposed to lead through drinking water pipes faced a risk of which they were unaware until the problem was discovered, but whose distribution was well-known. Exposure to lead in drinking water has been long-known to have serious health effects[9] and we know how to stop exposure (replace the pipes). What was unknown was (1) why the problem was not addressed earlier and (2) who would pay for fixing the problem and compensating residents. That is, the Flint lead exposure problem was that the residents were exposed without their knowledge by the municipal water system to a well-understood risk, not that there was a great deal of uncertainty about the problem, what was needed to solve it, or the extent of the harm. Regulators failed to take appropriate actions but their failure was not failing to understand the problem or solution.

Large-number problems involving changes in risk distributions are categorically different from individual problems. When risk distributions change, relying on responses by regulators and juries becomes problematic. Regulators struggle with tradeoffs and are slow to act or change course when new information arrives; juries struggle with evaluating whether the behavior was negligent. These situations need the data to understand whether there is a new risk distribution and, if so, what it is.[10] Economist Terry Anderson concisely summarized this: "The extent to which human beings react to climate change depends critically on the quantity and quality of information they have about the consequences."[11] As I describe below, ARTMs excel at incentivizing the development of such data.[12]

We need these incentives as existing regulatory solutions seem inadequate to address the consequences predicted by the current models. Carbon taxes, proposed or actual, are well below the levels necessary to reduce emissions by the amounts the models predict are needed.[13] Even the levels imposed are unpopular.[14] Governments of major sources of emissions are unwilling to reduce their emissions.[15] Emission reduction proposals for significant sources are ignored by some of the worst offenders.[16] Even the European Union, nominally committed to significant reductions, struggled to come to a consensus on a definition of "green finance," largely over whether to include natural gas and nuclear power production.[17] The Glasgow COP21 conference ended with the UK minister who organized it in tears over the watering down of commitments to reduce coal emissions.[18] However much future emissions are reduced, it does not appear they will be reduced enough to prevent significant impacts on climate. Hence, adaptation is an essential part of addressing climate change.

Central to my argument is the role of entrepreneurs. In Austrian economic theory, which offers the most developed account of their role, entrepreneurs identify opportunities to more efficiently use resources. These may be addressing a known need at a lower cost or meeting a previously overlooked or new need. Entrepreneurs' identification of such opportunities is their central function and requires knowledge. ARTMs provide a vehicle to deliver that knowledge because ARTMs are dependent on developing new sources, new modeling techniques, and other methods of better understanding the climate.

The chapter begins with the challenges posed by changes in risk distributions. It then turns to how ARTMs might address these challenges, discussing the development of ARTMs to date and the institutional requirements for their development. Finally, it concludes by suggesting how policymakers might best harness the ARTM market to address climate risks.

Challenges Presented by Changes in Risk Distributions

As Jonathan Adler noted, "Climate change is ultimately a problem about environmental risk, albeit a very large risk."[19] A warmer climate produces significantly different distributions of climate-related risks than a cooler one.[20] Partly this is due to the wide range of estimates for important parameters in climate models. For example, the most recent IPCC report's range in its "Summary for Policy Makers" for its predictions is quite wide. For the "Near term, 2021–2040," the report's five scenarios in the "*Very likely* range" of temperature increases cover ranges of 0.5 (1 scenario) or 0.6 (4 scenarios) degrees Celsius, which represents error ranges of 0.3 degrees around the "best estimate" predictions, roughly ±33%.[21] For the "Mid-term, 2041–2060" estimates, the range is wider, with scenarios predicting ranges from 0.8 to 1.1 °C. For the "Long term, 2081–2100" they are between 0.8 and 2.5 °C.[22] The report predicts "discernible increases" and "discernible changes" in a variety of weather effects from each 0.5-degree Celsius increase. These wide error bands around the best estimate predictions leave us with substantial uncertainty and so major challenges in modeling.

Models' ability to predict specific impacts (e.g., hurricanes off the Atlantic coast of North America, droughts in sub-Saharan Africa) is also subject to considerable uncertainty. Because many of the weather impacts are themselves subject to considerable uncertainties, even the IPCC's relatively strong level of confidence about, for example, predictions of increased heavy precipitation ("It is *very likely* that heavy precipitation events will intensify and become more frequent in most regions with additional global warming."[23]) leaves uncertainty as to how much and where the additional precipitation is going to fall.

This uncertainty means that the threats from climate change may represent changes in the distributions that go beyond simple shifts within a distribution. As an example, consider rainfall, often modeled using the Weibull distribution.[24] This distribution has two key parameters: the shape parameter (k) and the scale parameter (λ). These parameters affect the distribution's shape as depicted in Figs. 1 and 2 giving the distribution dramatically different appearance for two arbitrarily chosen values of the parameters.

As is evident, these different values produce radically differently shaped distributions, and predictions of rainfall using this distribution will vary considerably depending on the value of those parameters. For example, hypothetically assume the current rainfall distribution at a particular location follows a Weibull distribution with $\lambda = 0.5$ and $k = 2$ of Fig. 1.[25] Even if the Weibull distribution remains appropriate for rainfall with +1.5 °C, if the parameter values have shifted to $\lambda = 1.5$ and $k = 3$, as in Fig. 2, the shape of the distribution will be dramatically

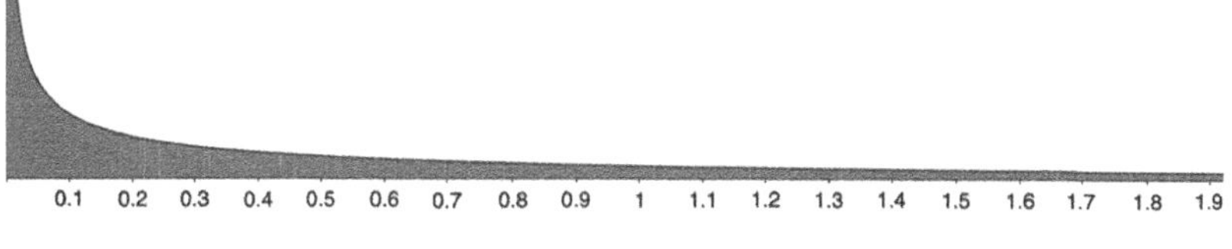

Fig. 1 Weibull distribution (0.5, 2)

different in terms of the total size of the tail of extreme events and the extent to which the far right of the tail extends.[26]

Of course, rather than merely producing a different shape to a distribution, climate change may change the distribution itself. The gamma distribution is also sometimes used to model rainfall. Figures 3 and 4 show gamma distributions with the same shape and scale parameters used in Figs. 1 and 2 (again, arbitrarily chosen for illustration purposes).

Predicting outcomes when we do not know the underlying probability distribution is even more challenging than it is when there is a reasonable hypothesis about the shape of that distribution based on historical data.

Given the uncertainties inherent in modeling something as complex as climate, it is likely that even where models are relatively certain in making high-level predictions, the degree of uncertainty is much greater in predicting, for example, changes in rainfall in Brazil's Paraná State, where forty years of data proved insufficient to choose between the Weibull and gamma distributions to characterize the existing distribution of rainfall in the article previously cited. Similarly, with hurricanes, the decision of whether to equally weigh historical data or give more

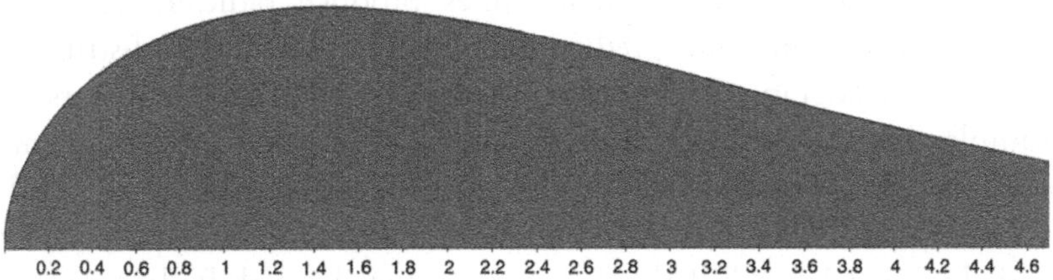

Fig. 2 Weibull distribution (1.5, 3)

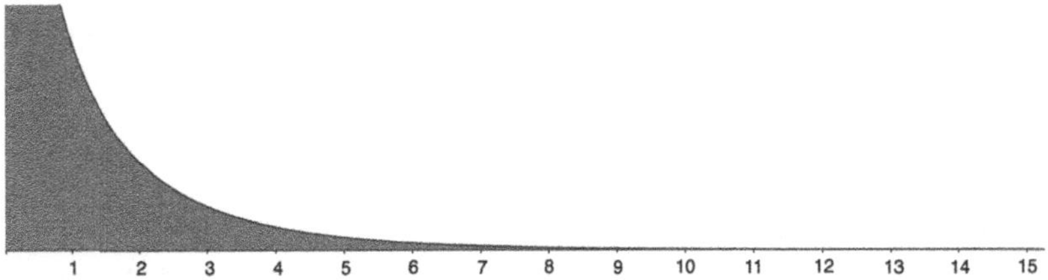

Fig. 3 Gamma distribution (0.5, 2)

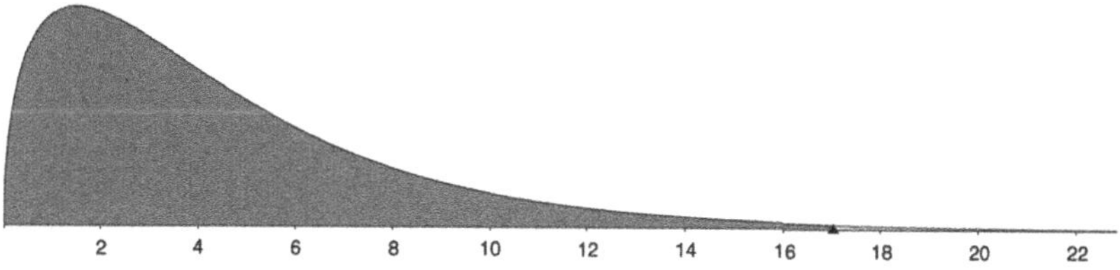

Fig. 4 Gamma distribution (1.5, 3)

weight to more recent data determines which of the two distributions performs better in back-testing. The model cannot tell us which distribution to use since it cannot solve the problem of determining whether there has been a regime switch.[27] Further, a possible regime switch affects the choices of both parameters and distribution.[28]

Firms that insure/reinsure catastrophic risks have experienced coping with difficult risk distribution issues.[29] Indeed, the science of modeling extreme events has made considerable progress in understanding such risks.[30] As the problems of coping with hard-to-model risks are something the insurance industry has been grappling with, we need not despair at meeting these challenges. Indeed, the growth of ARTMs to cope with catastrophic events has prompted investment in developing the tools being used to cope with these.

It is likely some important climate changes are not based on continuous functions, so discontinuous changes in weather features are likely consequences of changes in climate.[31] Benoit Mandelbrot made a series of theoretically devastating (although largely ignored) attacks on finance theory by pointing out that financial markets are not appropriately modeled with a normal distribution of error terms, as virtually all finance models do.[32] One of Mandelbrot's key points derived from the existence of moments when he found that trading time "speeds up" and prices move discontinuously without going through intermediate values. In these circumstances, a standing order to sell a stock bought at $40 when the price hits $30 is of no use as the price is never $30.[33] Something similar seems possible, and even likely, for climate change.[34] If, for example, the Gulf Stream were to collapse, as may result from climate change, the effects would be discontinuous.[35]

Because predicting such changes is difficult, the inevitable-but-unknown deficiencies in models likely mean they are underpredicted. For example, Boers' "observation-based" analysis of the Gulf Stream raises this point, noting that current models may make the Gulf Stream "too stable."[36] Overall, assessing the risks of catastrophic events requires answering difficult questions, which current data and modeling often cannot do at a high level of confidence: "Which catastrophe model is most appropriate for a specific type of risk exposure? How different are the results of different models? Are there known biases in some models related to specific perils or geographical regions? Are models for one region more credible than for another? How can we quantify the additional uncertainty related to the lower credibility of some models? Are there ways to validate some modelling results? What are the primary sources of uncertainty in the modelling?"[37] As Alex Krutov notes, "the list of questions never ends."[38]

Just because problems are hard does not mean they cannot be addressed by developing new techniques and data. A classic paper in extreme value modeling—inspired by floods that killed over 1800 people in the Netherlands, entitled *Fighting the arch enemy with mathematics*,[39] prompted significant progress in both the theoretical approach to extreme events and answering the practical question of the necessary height for dikes.[40]

Resolving these questions only begins the problem of translating predictions about the weather into data that can be acted on by firms and individuals. As Anderson notes, raw data is insufficient to drive adaptation, as we need to understand the interactions between natural, engineered, and economic systems. "Only by accounting for all three systems ... will the information be available to allow asset owners, financial institutions, and risk arbitrageurs to price resources, products, and services that will incentivize people to adapt."[41]

We face major challenges in creating actionable information to enable adaptation to change climate. The point is not simply that climate modeling is hard, that our models are likely flawed in predicting important climate phenomena, or that converting the output of models into information that can be used is much more complex than simply predicting future weather events, but that improving models to the point that they help economic actors make decisions will take a considerable investment in gathering and analyzing data. Only then can actors determine how to prevent or mitigate costs.

Such problems involve extremely large numbers of people. A major change in the Gulf Stream would dramatically affect (at a minimum) the eastern seaboard of the United States, where roughly a third of the population lives, even without considering the substantial impacts on populations further from the coast and outside North America. This is a different category of harm than the Flint water contamination, which involves "only" 100,000 people.[42] We have mechanisms to address Flint-scale environmental problems. However, existing legal tools are less capable of coping with the scale of problems predicted by climate change. In theory, some regulatory measures might, but the record of such efforts thus far is less than impressive[43] and even governments rhetorically committed to doing so seem to have chosen deadlines for their implementation well in the future, past the time horizon for current officeholders.[44]

Most attempts to harness markets to address climate change have disappointed so far. The reason is that the economic tools deployed or proposed (e.g., carbon taxes) often stem from models that offer no mechanism for errors to be corrected or new knowledge applied to problems. This is not surprising, as, for most economists today, "[c]hoice ... has come to mean the solution of a maximization problem."[45] By shifting to an alternative approach, I focus on how when we recognize that "entrepreneurship is *at the very heart* of the economic process,"[46] we can harness it to address climate change. These shifts focus from specifying the parameters of a maximization problem and its constraints to facilitating entrepreneurship. For entrepreneurs to succeed in adapting to problems posed by climate change requires developing the information necessary to understand its impacts, putting it into actionable form to allow firms and individuals to react to it, and reducing the transaction costs involved.

The Role of Alternative Risk Transfer Mechanisms

Risk transfer is a crucial element in determining whether a transaction or contract is treated by the legal system as insurance or not.[47] The general rule for which risks to transfer has long been that "core risks—those that are central to a firm's daily business—should be retained, while non-core risks—those that are a byproduct of daily business—should be transferred or hedged. The premise is that a company has information and expertise relating to its core risks and, therefore, greater ability to manage its exposures intelligently (e.g., safely, efficiently, and cost-effectively)."[48]

ARTMs blur the distinction between core and non-core risks. Once we consider adaptation, weather-related risks that might previously have been considered non-core become at least partially core risks. For example, a firm growing corn in the American Midwest faces risks from the historic levels of weather variations (temperature, rainfall).[49] The firm can do little to affect these. Hedging their impact reflects an appropriate shifting of a non-core risk. But if climate change is altering the local weather pattern such that changing to more drought-tolerant crops or moving production elsewhere is appropriate, climate-change-related shifts in weather patterns begin to look more like core risks.[50]

In some ways, the core/non-core distinction can be thought of as information is driven: "Those factors about which a firm perceives itself as having some comparative informational advantage will be those factors on which the business concentrates for its core business cash flows. Risks about which the firm has comparatively less information will be those risks more likely to be hedged, diversified away, insured, or controlled in some other fashion."[51] The agricultural producer in our example needs to learn whether weather fluctuations this year reflect a shift in the distribution of weather conditions or are a normal variation (or some combination) to decide whether changes in crops grown are needed for next year.

ARTMs can help discover how we can develop a better understanding of the consequences of climate change and how to best adapt to them. In the example, the agricultural firm is well-positioned to experiment with changes in crop varieties, planting locations, growing techniques, and a host of other decisions which represent margins along which it might adapt to a shift in weather patterns over and above historic levels of variation. Developing a mechanism by which it learns the impact of various risks on its operation—the consequence of using ARTMs—is crucial. Because ARTMs can be successful only when they are part of "a robust risk management process,"[52] firms using them must participate in that process.

The Traditional Insurance Baseline

We begin with a simple insurance transaction. Typically, an insured facing a risk will purchase insurance from an insurance company, which charges a premium based on the insurer's assessment of the probability of possible losses. The insurer (in this role, the cedent) will frequently reinsure a portion of risk to a reinsurer. The reinsurer may then cede some of the risks it assumed to other reinsurers. At each level, the insured/cedent may retain a portion of the risk, either because it wishes to do so or because the (re)insurer will insist (to motivate the insured to take precautions against incurring losses). Risks begin with the insured, move in part to the insurer, and then move in part to reinsurers.

This sort of routine insurance transaction handles many risks. Automobile liability insurance, for example, is typically handled by individuals buying a policy covering their automobiles, retaining a portion of the risk through the deductible, and shifting the remainder, subject to policy limits, to the insurer. The insurer may then cede a portion of its risks to a reinsurer. For fleet operators, there are more options. They can purchase policies covering fleets from a commercial insurer but may also opt to create a single-parent captive insurance company (an insurer they own), which insures the fleet. While this appears in some respects like the routine insurance transaction—there will be a policy written by a legal entity with its own legal personality to the insured, and the insurer will likely seek reinsurance—it has a crucial difference: the cost of the risk retained by the captive insurer will ultimately fall on the insured entity which owns the captive. This economic identity between the insured and the insurer makes a captive solution different from a standard insurance purchase, as the insured now has a larger interest in reducing the risk it has insured through its captive than it would in reducing an equivalent risk insured with a commercial insurer. In addition, putting the risk into a separate business unit within a firm brings organizational attention to the risks that the business unit handles. This incentivizes the development of new information by the business unit and entrepreneurial responses to the risks transferred to it.

A critical difference between the ARTMs discussed below and traditional insurance is that ARTMs are regulated differently. Regulation is an important factor in shaping insurance markets because "insurance is one of the most heavily regulated industries, a fact that, by itself, introduces a broad set of constraints and risks not found in other sectors."[53] Moreover, this regulation "is not uniform among jurisdictions, contributing to the fragmented nature of the insurance marketplace."[54] This creates an opportunity for specialist jurisdictions providing appropriate regulation to play a role, as discussed below. ARTM markets are where innovations develop in part because of the implicit tax imposed by regulation on conventional insurance markets.[55] We now turn to an exploration of how a range of ARTMs could change firms' reactions to climate change.

The Role of Alternative Risk Transfer

The market for ARTMs is constantly evolving, as entrepreneurs create new means of addressing risks. ARTMs are "'alternative' because they pierce the boundaries of conventional risk management concepts and techniques (e.g., pure insurance, reinsurance, derivatives), call on diverse financial engineering mechanisms from a number of different sectors, and/or draw in capital from a broad range of sources."[56] We consider three types of ARTMs, whose characteristics address climate change in different ways: captive insurance, catastrophe bonds, and weather derivatives.

Managing risk requires capital. Some ARTMs (captives) rely substantially or entirely on firms investing their own capital. Others (catastrophe bonds) require capital from outside investors. Still, others transfer risk without regard to whether the purchaser suffers a loss (weather derivatives), avoiding the need for an insurable interest or proof of loss.

Captive Insurance

There are a range of captive insurance structures possible today. We focus on the simplest, the single-parent captive.[57] The key features are that captives are owned by the insured; exist as a separate legal entity; design their own coverage; have access to the reinsurance market; and are domiciled in a jurisdiction with a regulator accustomed to captives.

Insureds' ownership of captives is critical to captives' ability to improve their owners' understanding of climate risk; it is in firms' capacity as captive owners that they are incentivized to learn about the risks and develop mitigation methods. Similarly, the captive's existence as a separate legal entity,[58] its regulation by its domicile, and reinsurers' evaluation of it as a potential cedant forces investment in the evaluation of the risks being insured by its owner, its regulators (and the various service providers licensed by the regulator who must regularly produce accounting and other reports on the captive), and reinsurers.[59]

When the captive is formed, it must convince its regulator it has properly evaluated the risks it is insuring and is sufficiently capitalized to provide the coverage it is writing. It must then regularly demonstrate to its regulator that its plan is succeeding, requiring investing in data on both risks and efforts to reduce them.[60] To effectively manage its risks, of course, a firm must understand those risks. Requiring firms to develop this understanding is a key benefit of captives.[61]

The final important characteristic of captive insurance for our purposes is the relationship between captives and their domiciles. Most captives are domiciled in one of a few jurisdictions. (In the United States, Vermont, the District of Columbia, South Carolina, and Tennessee are major U.S. captive domiciles; U.S. captive owners also make use of foreign domiciles, such as Bermuda.) Captives choose these jurisdictions because they successfully developed specific captives regulatory regimes.[62] For example, captives generally are not subject to the kind of pre-approval regulation of policy language that virtually all U.S. states impose on insurers selling to the general public.[63] Because captives are insuring their owners in most cases, the usual consumer protection rationale for such regulation does not exist.[64] Indeed, some firms use the creation of a captive as a signal to the market that they take particular risks seriously.[65]

This is not the only way in which captive domiciles' regulatory regimes differ, however. As Charlotte Ku and I have argued elsewhere, jurisdictions that are successful as captive domiciles (and as financial centers generally) are successful because they invest in good governance, in keeping their statutes and regulatory regimes up-to-date, and in creating a strong regulatory community.[66] It is crucial to the success of these jurisdictions that captives domiciled in them succeed—the reputational blackeye from problematic captives would drive out the successful players.[67] Appropriate regulation, not laxness, makes a successful jurisdiction. Indeed, at least in part, firms' desire for a captive domicile with serious and appropriate regulation is driven by their need to keep undesirable players out of the domicile.

The importance of a domicile being able to accommodate innovation can be seen in Bermuda's response to the insurance crisis prompted by the extraordinarily destructive 1992 hurricane season. Bermuda developed the regulatory capacity to become a reinsurance center in response to the increased demand for capital by a global reinsurance industry reeling from devastating losses.[68] The resulting "Class of 1993" specialist reinsurers dealing with catastrophes represented a major increase in capacity and would not have been possible without Bermuda's regulatory agility. As Hartwig and Wilkinson note, "Bermuda's insurance regulation is designed to facilitate the creation of companies and insurance products while ensuring the companies operate responsibly within specific margins of solvency."[69]

Captives thus offer an ARTM that provides clear incentives for firms to develop information on the risks they face and adopt means of mitigating the harms if the risks materialize or avoid the risks. The obstacles to using captives come in part from the relatively substantial fixed set-up costs and from regulatory obstacles. The former can be reduced by the development of means of sharing fixed costs (e.g., "rent-a-captives" and risk retention groups). The latter is constrained by the competition among jurisdictions to be captive-friendly domiciles.[70]

Catastrophe Bonds

Catastrophe bonds are a sub-class of insurance-linked securities (ILS) whose risk is based on potential catastrophes. ILS are "financial instruments, other than traditional equity and debt securities issued by insurance companies, which carry insurance risk or a type of risk that is closely related to it."[71] They developed in response to the losses caused by Hurricane Hugo in 1989 and in the 1992 hurricane season, although what would likely have been the first catastrophe bond was planned before 1992's Hurricane Andrew and postponed because of it.[72] The experience made clear the limits to the catastrophic risk that any insurer or reinsurer could assume. The solution was to turn to the markets for capital infusions in new forms. "More than any other product it [trade in alternative products after Andrew] turned ILS into one of the reinsurance industry's greatest recent innovations."[73]

These bonds are usually structured through a special purpose reinsurer (SPR), a legal entity distinct from the firm arranging the transaction; the SPR issues notes to investors, the proceeds of which are invested in safe assets. The cedant pays a premium to the SPR. The invested collateral is used to repay the principal to the investors at maturity, unless the specified catastrophe triggers the payout of some or all of the principal to the cedant, which has thus mitigated its exposure to the catastrophic event.[74] The cedant is paid when a bond is triggered, typically by a combination of a weather event (e.g., a hurricane of pre-defined force striking pre-defined locations within a pre-defined window of time) and a level of loss (e.g., total claims for all insurers for the pre-defined region reaching a pre-defined level). Among the requirements is a jurisdiction that can be counted on to support the SPR in distributing the assets as the bond's terms require, respecting the bankruptcy-remoteness of the SPR from the organizer, and reducing the transaction costs of organizing the SPR.

Because they are triggered by weather events, catastrophe bonds require extensive modeling to design.[75] While the data generally comes from public sources, different sources' models differ.[76] Moreover, the models take into account not just the cause of the catastrophe (e.g., the link between climate and hurricanes) but also the characteristics of the property insured. Experience is improving these models.[77] As noted earlier, a significant challenge is determining when past data is/is not a reliable basis for a current model.[78]

The market has "evolved significantly" since ILS appeared, with "a solid, expanding core of experienced investors" and important roles played by "[r]ating agencies and cat modeling firms … in increasing the confidence of market participants by working on and providing analysis of nearly all transactions."[79] In addition, issuers are more comfortable with ILS, which allowed their use to evolve "from being somewhat experimental to an integrated part of the risk management toolkit."[80] Their structure also developed: bonds increasingly involve multiple perils, multiple years, and multiple tranches.[81] Bonds have expanded to non-catastrophic risks, such as triggers based on temperature indices in 19 U.S. cities.[82] Most importantly, as the term lengthens on catastrophe bonds, longer-term climate impacts become more relevant to their evaluation, increasing their usefulness in incentivizing better modeling of medium-term climate impacts. A related product, reinsurance sidecars, "made a sudden appearance in the aftermath of the 2005 Katrina-Rita-Wilma hurricane season; they addressed an urgent need and then quietly decreased in importance."[83] The market's ability to try innovations and discard those that are less desirable is an important element in the superiority of market-based methods of addressing problems.[84]

Catastrophe bonds, and ILS generally, offer three advantages for climate risk. First, structuring and evaluating the bonds requires the development of ever-more-sophisticated models of the risks included. For example, the risks embedded in a seven-year bond for hurricanes in Florida can be understood only if the sensitivity of the model used to design its triggers is comprehensible to buyers and sellers and if both have confidence in the data underlying the design of the trigger. *Ex ante,* both have an interest in making it possible for both sides to evaluate the bond's risks.[85] Second, they work only if there are active participants on both sides of the market.[86] Those assuming the climate risk acquire, along with their weather-related ILS, an immediate interest in seeing mitigation measures taken by public and private entities, to reduce the likelihood of the trigger's activation. Third, bonds are tradeable. If there is an active market, when a market participant learns new information, there is an opportunity for a profitable trade. The existence of a market thus incentivizes investors to engage in the discovery of new information on climate risks.[87]

Weather Derivatives

Demand for weather derivatives comes because the weather market includes "natural hedgers on both sides."[88] Indeed, "[t]emperature variations, climate fluctuations, and weather fronts mean money. Lots of money."[89] After a somewhat rocky start, there is an established market for weather and catastrophe derivatives, although opportunities for improvements remain.[90] As with catastrophe bonds, a key to weather derivatives' growth is improved modeling[91] and the greater number of weather derivatives and the construction of portfolios requiring hedging of weather risk increases demand for further modeling improvements.[92]

Weather derivatives may be structured in different ways. For example, "[t]he payout may depend on a hurricane of specific magnitude making a landfall in a certain area; on the value of total cumulative losses from hurricanes to the insurance industry over a certain period of time for a specified geographical region; or on the value of an index tracking the severity of an earthquake at several locations."[93] Derivatives can be customized and traded over the counter or they can be based on standardized forms and exchange traded.[94]

Weather derivatives include those tied to heating and cooling degree days, calculated based on measurement of temperatures at specified locations and generally based on government-supplied data.[95] These are often used to hedge energy demand and in agriculture, tourism, retail, and construction.[96] Rainfall, snow, wind speed, and sunshine hours also serve as the basis for derivative construction.[97] In an innovative use, the World Food Programme (WFP) bought weather derivatives on drought conditions in Ethiopia, based on precipitation measured at multiple locations converted into an aggregated crop stress index. If conditions met those specified, the payout provided immediate access to funding. Fortunately, conditions were not severe enough to trigger the payout.[98]

Weather derivatives open up ARTMs to sectors other than those with regular insurance needs or insurers seeking additional capital. Particularly as the exchange-traded market grows, they broaden access to opportunities to hedge climate risks. As with the other ARTMs, weather derivatives benefit from improved modeling and data. Both buyers and sellers need to understand the derivatives' construction. The WFP example demonstrates how the use of weather derivatives can be both useful (the WFP acquired access to funds if it needed them at a fraction of amount that would be needed) and drive knowledge development (the construction of the crop stress index for Ethiopia will enable further transactions and advances knowledge of droughts).

Using ART Methods

The theme connecting these ARTMs is their reliance on developing data and improving our modeling capacity, enabling predicting the impacts of climate change on more granular levels. Armed with reliable predictions, firms and individuals can better adapt to new conditions.[99] However, even if we develop better models and more data, the task is not complete because "[i]f the modelling software is a complete black box to an investor, any analysis of its output is limited and deficient."[100] An important part of developing the market for ARTMs will be to continue to invest in unpacking that risk analysis black box, which itself will educate market participants about climate risks.

If "one of the areas most ripe for innovation" is "the development of novel financial instruments,"[101] how do we get more of this innovation? Who will develop the new models and data we need? Entrepreneurs. The ARTs described above all contain important elements of entrepreneurship in creating and acting upon solutions. To explore, we need insight into what entrepreneurs do. Perhaps no modern economist has devoted more effort to studying entrepreneurship than Israel Kirzner.[102] Working in the Austrian economics tradition, in his view, the entrepreneur's function is to engage in "the correction of 'maladjustments' in market prices and decisions."[103] Entrepreneurs accomplish this by being alert to unnoticed opportunities.[104] This is what is needed to improve our ability to adapt to climate change.[105] For example, after Hurricane Katrina and the 2004–2005 hurricane season,

> there was an almost universal conviction that the frequency of hurricanes in the widely used commercial models was significantly understated. (There were also concerns about how other modules of the models performed, and whether the damage and loss severity were understated.) Since then, the models have been modified to produce loss results that are greater than would be expected based purely on long-term historical data, either as the main output or as an option available to the user. The change reflects the view that the long-term observations do not represent the current atmospheric conditions that affect formation, development and landfalling of tropical storms and hurricanes.[106]

This is a near-ideal example of correcting maladjustment in market prices and decisions. In the next section, I apply Kirzner's insights to how ARTMs may be harnessed to improve our ability to adapt to climate change and potential policies that encourage the application of ARTMs to problems related to climate change.

Harnessing ARTMs to Incentivize Adaptation

Harnessing ARTMs to address the challenges posed by climate change requires understanding the role entrepreneurs can play in developing and deploying solutions. To do that, we need to explore the entrepreneurial function and then find the obstacles that hinder the development of solutions. I begin with Kirzner's insights into the role of entrepreneurs and then apply those insights to the ARTMs discussed above.

The Entrepreneur's Role

The insight that Ludwig von Mises, Friedrich Hayek, and the Austrian School generally brought to the study of entrepreneurship was "an appreciation for the market process through which the scattered information, concerning resource availabilities and consumer valuations, is mobilized and brought to bear upon the decisions governing the production and the allocation of resources."[107] Austrian entrepreneurship theory sees the market process as "a systematic process of knowledge expansion, the equilibrating character of which is the expression of entrepreneurial discovery."[108] This insight is critical for harnessing the power of markets to address problems like climate change. As Kirzner noted, if we neglect the Austrian view of entrepreneurship (and subjectivism more generally), we learn nothing about the process of adjustment, precisely the most critical process here.[109]

In Kirznerian entrepreneurship theory, the entrepreneur's role is to discover market errors and act upon that knowledge.[110] The entrepreneur acts in the market by competing with those who have not seen those errors. This focuses on the process by which entrepreneurs act in the marketplace. By contrast, neoclassical economics treats competition as "as *a state of affairs,* most completely captured by the notion of 'perfect competition'" where economic actors "are assumed to be confronted by prices (for products or resource services) at which each is free to buy or to sell whatever he chooses, in whatever quantities he pleases, without having to worry about not being able to carry out his buying and selling plans successfully, and with assurance that the respective prices will not depend on his own buying or selling plan."[111] In the Austrian view, on the other hand, competition involves "the dynamic thrust and counter-thrust of an active market process." Competition is "a process of give-and-take during which prices are continually buffeted by entrepreneurial impetuosity and activity."[112] Thus, the market is not an institution charged with computing the solution to the problem of allocating resources to their highest and best use but one "that is remarkably effective in *discovering* where existing resource utilization falls short of the 'best'—most valuable—pattern conceivable."[113] At any given time in the market "ghastly errors are doubtless being made in allocating resources."[114] What the market enables is the discovery of those errors and the creation of means to replace them with "less faulty" patterns of resource use.[115]

An example of this process in action is how United Grain Growers, a Manitoba-based firm, reduced the risks it faced from price volatility.[116] Even after hedging currency and commodity price risks and buying property and liability insurance, United believed its earnings still had too much volatility. It found the remaining volatility in changes in the volume of grain sold, which was heavily affected by weather. The firm created a financial contract that paid United if *industry* volume fell below a specified level. This example demonstrates—in a relatively simple case—how solutions are found by first developing a deeper understanding of the situation (discovering that weather drove significant earnings volatility) and then creatively finding a means to correct what had been a "ghastly error" and replace it with a "less faulty" approach.

Another example is the evolution of fire insurance. After struggling to accurately underwrite these risks, the insurance industry developed the "Universal Mercantile System" in 1893, which rated buildings and towns, developed a standard building for each town, and then adjusted rates based on comparison with the standard: "building codes, fire department capabilities, water supply, and the efficiency of the police department. Credits and deductions were made if the actual elements were better or worse than the standard. Another set of credits and deductions were made based on certain building features, such as area, construction, occupancies and adjacent exposures."[117] Again, the entrepreneurial response was to invest in the information necessary to enable a solution.

More generally, we can see this process at work in the ARTM market. The primary drivers of the market have been "(1) insufficient capacity of the insurance and reinsurance industry to withstand the impact of catastrophic events without transferring some of the catastrophe risk to investors; (2) accounting rules that make it more capital-efficient to transfer some of the insurance risks to the capital markets; and (3) growing investor demand for assets that have low correlation with the traditional financial markets."[118] The ARTM market is shifting climate risk decisions to risk managers, who have a "comparative advantage in coping with new risks helping align incentive structures."[119] That in turn will "encourage the development of new risk transfer tools and accelerate efficient adaptation."[120]

This process of discovery is not simple. Entrepreneurship is not something "exercised flawlessly, tirelessly, and universally" but "a scarce, valuable resource" of which we need to "begin to take careful account."[121] If we are to encourage exercise of the entrepreneurial function of noticing "what people have overlooked"[122] with respect to climate change, we need to begin with encouraging the development of data and models. Unfortunately, many regulatory programs are crafted without awareness of the market's discovery process and so do not leverage its capacity for generating knowledge.[123] One of the most serious consequences of that lack of awareness is when regulatory measures "block activities that have not yet been foreseen by anyone, including the regulatory authorities."[124] This is precisely the obstacle to deploying ARTMs to address climate change and so we now turn to examine the type of policy options that might facilitate entrepreneurs finding ways to use them.

Policy Options

To harness ARTMs to adaptation to climate change, we need to—as Anderson frequently prescribes—put on our "Coase-colored glasses."[125] This means that we need to focus on finding ways to reduce transaction costs to facilitate market adjustments. As Anderson observes, "[a] Coasian focus opens the door for entrepreneurs who see opportunities available from creating new property rights where they are lacking or contracting over existing property rights to improve resource use. Whether it is new property rights or new contracts, the entrepreneur captures previously dissipated rents."[126] When Kirznerian entrepreneurs are viewed through Coase-colored glasses, we can identify potential policy solutions that remove obstacles from their paths.

Wearing Coase-colored glasses to harness ARTMs to address the challenges posed by climate change is important because regulation has kept the insurance industry from being as innovative as it could be.[127] For example, because of the inhibiting effects of regulation, the industry responded well to post-World War II growth in demand from those seeking to transfer traditional risks but did not respond as well to new risks. As one long-time industry participant explained in 1984, "[t]he insurance industry continued to use 18th, 19th, and early 20th century insuring agreements, exclusions and conditions for late 20th century exposures."[128] Eventually, as that same observer predicted and despite the regulation-induced slowing of innovation, the development of captives, "new and varied risk financing plans," a greater emphasis on loss control, and other techniques developed to handle these risks.[129] Avoiding repeating those mistakes and providing public goods such as supporting data collection and research into model development are straightforward first steps.[130]

Since the 1980s regulators and legislators have taken important steps to spur innovation in insurance. For example, the Gramm-Leach-Bliley Act in 1999 eliminated barriers to banks and investment banks intruding into insurance companies' territory (and vice versa), expanding the range of potential entrepreneurs developing solutions and allowing a combination of skills across industries.[131] One example is the impact on insurance companies and investment banks, which traditionally used quite different techniques to price products. As the two industries converge, "the opportunities for considering alternative pricing approaches are growing (e.g., use of simulation processes in insurance pricing, the application of extreme value theory in examining low-frequency financial and non-financial catastrophes, and so forth)."[132] In addition, as hedge funds, themselves innovative forces in finance, become important providers of capital for catastrophe and weather risk, they spur innovation through demand for new financial instruments.[133] A parallel convergence between corporate finance and risk management—each of which previously "had its own definitions of risk, its own transactions to help firms manage risk, its own special skill sets for risk analysis (actuarial analysis in insurance versus econometrics and stochastic calculus in derivatives), and its own vocabulary for selling risk management to corporate clients"[134]—is underway, a process which Christopher Culp termed "a convergence in a way of thinking."[135]

One dramatic development has been the captive insurance industry. As I have elaborated elsewhere, considerable innovation in insurance-based solutions, including captives, has been unleashed in U.S. markets by the federal government's creation of non-territorial special economic zones in insurance.[136] By adopting a "lead state regulator" model for a particular type of captive limited to liability insurance (risk retention groups), blocking the application of state laws forbidding "fictious groups" from forming to purchase insurance, and standardizing the conditions to participate in the "surplus lines" insurance market (which permits non-admitted carriers to write insurance where no admitted carrier does so), the federal government has used regulatory competition among the states to expand the range of insurance-based solutions for a number of insurance needs.[137]

Entrepreneurs used the opportunities created by these statutes' preemption of state regulations to innovate. For example, a group of structural engineers, who formed a purchasing group to buy liability insurance, developed their own "Technical Practice Review" process and ensured that the group's members were complying with professional guidelines and "actively employing effective loss prevention techniques."[138] While neither rocket science nor climate modeling illustrates how innovation can be incentivized by removing regulatory barriers to new entrants to a market: The structural engineers' group's "Coase-colored glasses" spotted an opportunity that the insurance market had not previously noticed or, if noticed, been able to exploit.

Many of these solutions could be easily adapted to address climate-related risks. The Liability Risk Retention Act (LRRA)[139] could be amended to permit risk retention groups and purchasing groups to provide climate-change-related insurance in addition to the liability insurance categories already provided. The Nonadmitted and Reinsurance Reform Act[140] could be amended to expand its streamlined process to climate-related products generally. The LRRA could be further amended to permit insurers from selected foreign jurisdictions to participate, something that its predecessor, the Products Liability Risk Retention Act, permitted but which was ended by the LRRA.[141] Such measures would harness the "law market" by expanding market pressures for states to innovate in regulating entities seeking to meet the needs identified in this chapter.[142] The size of the potential market for such solutions is considerable, with estimates of 20% of the U.S. economy being sensitive to weather conditions,[143] while the sophistication of many market participants remains relatively underdeveloped.[144]

Commenting on improving U.S. economic competitiveness generally, Kirzner argued that "[t]he way to wake up the U.S. is not to spend costly resources on expensive alarm clocks. It is to open the shutters and permit the light of opportunity to perform its own stimulation. Opening up the economy and eliminating restrictive regulations would stimulate alertness and compel existing firms to stay on their toes to forestall aggressive competition by others."[145] His prescription applies to adapting to climate change. Looking through our Coase-colored glasses, we can identify obstacles to developing ARTMs that will improve the data and the models that we use to understand climate and help us to connect that knowledge to the problems faced by individual actors. By embracing the price discovery process of markets, we can increase climate resilience.[146]

Notes

1. Thanks to Jonathan H. Adler, Roger E. Meiners, and Benjamin Simon for comments on earlier drafts.
2. J. David Cummings, "Property-Liability Insurance Price Deregulation: The Last Bastion?," in *Deregulating Property Liability-Insurance: Restoring Competition and Increasing Market Efficiency*, ed. J. David Cummings (Washington, DC: AEI-Brookings Joint Center for Regulatory Studies, 2002), 2.
3. "Pollution Liability Insurance—What Is Pollution Liability Insurance?" Zurich North America, accessed July 5, 2022, https://www.zurichna.com/insurance/pollution-liability.
4. Zurich's policies include claims assistance from the Zurich Environmental Emergency Response team, which provides a 24-hour call center, contractors to conduct cleanups, and coordination of regulatory reporting. "*Pollution Liability Insurance—Environmental Insurance Claims Assistance*," Zurich North America, accessed July 5, 2022, https://www.zurichna.com/insurance/pollution-liability.

5. Associated Press, "Jury Awards BP Workers $100 Million in Toxic Substance Case," *New York Times*, December 18, 2009.
6. Brett Clayton, "BP Wins $99 Million Reduction in Texas City Fumes Case," *Houston Chronicle*, March 16, 2010.
7. Mary Ann Azevedo, "Mass Suit Filed Against BP Over Alleged Texas City Benzene Release," *Houston Business Journal*, September 15, 2010.
8. *See* Charles Perrow, *Normal Accidents: Living with High-Risk Technologies* (Princeton, NJ: Princeton University Press, 1999 [1984]) (discussing inherent inadequacies of regulation of chemical plant operations).
9. Perri Zeitz Ruckart et al., "The Flint Water Crisis: A Coordinated Public Health Emergency Response and Recovery Initiative," *Journal of Public Health Management & Practice* 25 (2019): S84.
10. This idea is related to, but distinct from, Omri Ben-Shahar and Kyle Logue's proposal to address weather hazards through the incentives created by risk-adjusted insurance premiums. *See* Omri Ben-Shahar and Kyle D. Logue, "Perverse Incentives of Subsidized Weather Insurance," *Stanford Law Review* 68 (2016): 571. Among other differences, their primary focus is on using homeowner's policies to incentivize home purchasers to behave differently through price signals. Id. at 582.
11. Terry L. Anderson, "Introduction," in *Adapt or Be Adept: Market Responses to Climate Change*, ed. Terry L. Anderson (Stanford: Hoover Institution Press, 2021), 6.
12. The argument here is thus different from John Thrasher's elsewhere in this volume. Rather than climate change being a problem of irreducible uncertainty, as he posits, I argue addressing climate change depends on harnessing market mechanisms to reduce the uncertainty. John Thrasher, "Permission, Prohibition, and Dynamism," this volume. My argument aligns more closely with Mark Pennington's observations that markets are likely to be able to do better forecasting than states and that the problem with climate change is not market failure but the failure to establish markets. Mark Pennington, "Climate Change, Political Economy, and the Problem of Comparative Institutions Analysis," this volume.
13. Benjamin Zycher, "Hearken sinners: The end is near," *AEI* (October 22, 2018), accessed July 5, 2022, https://www.aei.org/articles/hearken-sinners-the-end-is-near/ (calculating that the midpoint estimate for the carbon tax necessary in 2030 that would achieve the goal of reducing

emissions would translate into a tax of over $29 per gallon of gasoline in 2010 dollars).

14. Stephanie Haynes, "Climate Conundrum: Tax on Emissions Is Pragmatic but Unpopular," *Christian Science Monitor*, July 15, 2021.
15. Ehan Masood and Jeff Tollefson, "'COP26 Hasn't Solved the Problem': Scientists React to UN Climate Deal," *The Guardian*, November 14, 2021.
16. Michael Standaert, "Despite Pledges to Cut Emissions, China Goes on a Coal Spree," *YaleEnvironment360*, March 24, 2021, https://e360.yale.edu/features/despite-pledges-to-cut-emissions-china-goes-on-a-coal-spree.
17. Mehreen Khan, "Scientists Lambast EU over Gas and Nuclear's 'Green' Energy Label," *Financial Times*, January 20, 2022. *See also* Catherine M. Sharkey, "Common Law Tort as a Transitional Regulatory Regime: A New Perspective on Climate Change Litigation," this volume (discussing inadequacies of many government plans for climate change which focus on the need for more research).
18. Kate Dennett, "Tears for a COP Out: Climate Conference President Alok Sharma Chokes Up and Tells Delegates He Is 'Deeply Sorry' as He Signs 'Watered Down' Green Deal After China and India Force a Late Change over Coal," *The Daily Mail*, November 13, 2021.
19. Jonathan H. Adler, "Water Marketing as an Adaptive Response to the Threat of Climate Change," *Hamline Law Review* 31 (2008): 730, 754.
20. As Alex Krutov explains,

    > The expectations of the future climate state are different from its current one. The effects of climate change relevant to hurricane activity, in particular the increase in sea-surface temperature, can already be observed. These changes make it harder to rely on the old approach of forming conclusions about future natural catastrophe activity based entirely on prior historical observations. The future frequency and severity of hurricane events might be a function of atmospheric and oceanic processes that are different from the ones in the period of historical observations.

    Alex Krutov, *Investing in Insurance Risk: Insurance-Linked Securities: A Practitioner's Perspective* (London: Risk Books, 2010), 111.
21. IPCC AR6 WGI, *Summary for Policymakers*, SPM-18.
22. IPCC *Summary*, SPM-18.
23. IPCC *Summary*, SPM-20.

24. Frederico Márcio Corrêa Vieira, Juliane Mayara Casarim Machado, Edgar de Souza Vismara and Jean Carlo Possenti, "Probability Distributions of Frequency Analysis of Rainfall at the Southwest Region of Paraná State, Brazil," *Revista de Ciências Agroveterinárias* 17, no. 2 (2018): 260, 261, 265 (The Weibull function is one of the most used functions in the field of climatology and meteorology" and concluding that it and the gamma function were the best fit for the rainfall data analyzed). There is considerable disagreement over the appropriate distribution to use for some weather phenomena. For example, hurricane models use three different distributions (Poisson, negative binomial, and binomial) and some commercial modeling software allows users to choose the distribution to apply. Krutov, *Practitioner's Perspective*, 89. *See also* Paul Embrechts, Claudia Klüppelberg, and Thomas Mikosch, *Modelling Extremal Events for Insurance and Finance* (New York: Springer, 1997), 9 ("in reality claims are mostly modelled by heavy-tailed distributions like Pareto, loggamma, lognormal, or heavy-tailed Weibull").
25. Note that these are completely hypothetical numbers merely to illustrate what it means to have a change in probability distributions whose extent is unknown.
26. Krutov, *Practitioner's Perspective*, 89 (choosing the parameters for a distribution is potentially "be a bigger issue than the analytical form of the distribution").
27. Krutov, *Practitioner's Perspective*, 90 ("if it [the distribution mean] is determined based on the full historical database of observations, with all observations given the same weight, negative binomial distribution seems to almost always outperform Poisson in back-testing regardless of the geographical region being considered").
28. Krutov, *Practitioner's Perspective*, 89.
29. In general, insurance loss distributions "are referred to as fat-tailed distributions because of the high probability of extreme diversion from the mean. (More precisely, these are leptokurtic distributions. Their excess kurtosis leads to the higher probability of outliers in a sample relative to samples drawn from a Gaussian distribution.)" Krutov, *Practitioner's Perspective*, 26; Richard H. Bernero, "Second-Generation OTC Derivatives and Structured Products: Catastrophe Bonds, Catastrophe Swaps, and Life Insurance Securitizations" in *Securitized Insurance Risk: Strategic Opportunities for Insurers and Investors*, ed. Michael Himick (Dearborn, Glenlake Publishing Company: 1998),

80 (describing improved risk modeling that resulted from Hurricane Andrew).

30. See, e.g., Embrechts, Klüppelberg, and Mikosch, *Modelling Extremal Events*.
31. National Research Council, *Abrupt Climate Change: Inevitable Surprises* (Washington, DC: National Research Council, 2002), 1 ("Abrupt climate changes were especially common when the climate system was forced to change most rapidly. Thus, greenhouse warming and other human alterations of the earth system may increase the possibility of large, abrupt, and unwelcome regional or global climatic events").
32. Benoit B. Mandelbrot, *Fractals and Scaling in Finance: Discontinuity, Concentration, and Risk* (New York: Springer, 1997) and Benoit B. Mandelbrot and Richard L. Hudson, *The (Mis)behavior of Markets: A Fractal View of Risk, Ruin, and Reward* (London: Profile Books, 2008) collect and popularize, respectively, Mandelbrot's analysis.
33. Mandelbrot and Hudson, *(Mis)behavior*, 22–23.
34. See, e.g., Krutov, *Practitioner's Perspective*, 89 ("When considering the choice of probability distribution for hurricane frequency, parameterization might be a bigger issue than the analytical form of the distribution. This is particularly challenging because of the varying views on the changes in hurricane frequencies over time. In fact, the regime switch view on the hurricane frequency affects both the choice of the parameters of the distribution and the choice of the distribution itself").
35. Niklas Boers, "Observation-based early-warning signals for a collapse of the Atlantic Meridional Overturning Circulation," *Nature Climate Change* 11 (August 2021): 680.
36. Boers, "Early-warning signals," 687.
37. Krutov, *Practitioner's Perspective*, 108.
38. Krutov, *Practitioner's Perspective*, 108.
39. L. de Haan, "Fighting the Arch-Enemy with Mathematics," *Statistica Neerlandica* 44 (1990): 45.
40. Embrechts, Klüppelberg, and Mikosch, *Modelling Extremal Events*, 288–289.
41. Anderson, "Introduction," 8. Some predictions are relatively easily translated into actionable form by markets, of course. For example, one study found that Miami-Dade County home prices rise with higher elevation and that beach front property there now has a 7% discount

(a discount that has been rising), suggesting that buyers are taking into account predictions of rising sea levels. Id. at 10.

42. CDC, "Flint Lead Exposure Registry," available at https://www.cdc.gov/nceh/lead/programs/flint-registry.htm.
43. For example, the decline in U.S. greenhouse gas emissions up to 2019 was largely due to the substitution of natural gas for coal and the economic slowdown due to the COVID-19 pandemic. As the economy rebounded and natural gas prices rise, U.S. greenhouse gas emissions rose. Andrew Freedman, "U.S. greenhouse gas emissions jumped in 2021," *Axios*, January 10, 2022; Brad Plummer, "U.S. Greenhouse Gas Emissions Bounced Back Sharply in 2021," *New York Times*, January 10, 2022.
44. For example, the British government of Prime Minister Boris Johnson announced bans on gas home heating and gasoline and diesel vehicles, both scheduled to take effect well after Prime Minister Johnson would be out of office even if he served the full five-year term of the current parliament and won a majority at the next election and served an additional full term (ignoring his present political difficulties). Dimitris Mavrokefalidis, "Boris Johnson Confirms Ban on All New Gas Boilers by 2035," *Energy Live News*, October 18, 2021, https://www.energylivenews.com/2021/10/18/boris-johnson-confirms-ban-on-all-new-gas-boilers-by-2035/; David Vetter, "Britain Will Ban Gasoline Cars in 2030. Why Are Experts Not Impressed?," *Forbes*, November 18, 2021.
45. Israel M. Kirzner, "Entrepreneurship, Choice, and Freedom," in *The Collected Works of Israel M. Kirzner: Reflection on Ethics, Freedom, Welfare Economics, Policy, and the Legacy of Austrian Economics*, eds. Peter J. Boettke and Frédéric Sautet (Indianapolis, IN: Liberty Fund, 2018), 4. Similarly, Kenneth Costello rightly criticizes much of the economic analysis of climate policy as what Ronald Coase termed "blackboard economics" which "abstract from the prevailing politics, the acquisition of tolerably accurate information on the social cost of carbon, the institutional setting, and the likelihood that the tax will achieve its objective. Because of politics, for example, a carbon tax is less likely to balance social and private benefits than to appease those in charge politically." Kenneth W. Costello, "The Political Realities of Climate Policy," in *Adapt or Be Adept*, 54.
46. Israel M. Kirzner, "Comment: X-Inefficiency, Error, and the Scope for Entrepreneurship," in *The Collected Works of Israel M. Kirzner: The Essence of Entrepreneurship and the Nature and Significance of Market*

Process, eds. Peter J. Boettke and Frédéric Sautet (Indianapolis, IN: Liberty Fund, 2018), 267.

47. *See* Helvering v. Le Gierse, 312 U.S. 531, 539 (1941) ("Historically and commonly, insurance involves risk shifting and risk distributing").
48. Erik Banks, *Alternative Risk Transfer: Integrated Risk Management Through Insurance, Reinsurance, and Capital Markets* (Chichester, UK: John Wiley & Sons Ltd., 2005), 10. Culp describes the difference between core and non-core risks:

    > The *core* risks facing a firm may be defined as those risks that the firm is in business to bear and manage so that it can earn returns in excess of the risk-free rate. *Noncore* risks, by contrast, are risks to which a firm's primary business exposes it to but that the firm does not necessarily need to retain in order to engage in its primary business line. The firm may well be exposed to noncore risks, but it may not wish to *remain* exposed to those risks. Core risks, by contrast, are those risks the firm is literally in business *not* to get rid of.

    Christopher L. Culp, *Structured Finance and Insurance: The ART of Managing Capital and Risk* (Hoboken, NJ: John Wiley & Sons, Inc., 2006), 30.
49. Firms that buy the crops, sell the growers supplies, or operate other businesses in the communities where growers operate are also exposed to these same risks.
50. It is, of course, possible that some consequences of climate change are so catastrophic as to make adaptation impossible in some instances. The assumption here is that, as is most often the case, changes on margins are the most important way to adapt. As we are locked into considerable durable capital stock, and because much of the greenhouse gases that have already been emitted will be in the atmosphere for decades, a focus on increasing the speed of adaptation will be critical in many instances. *See* Gregory W. Characklis, Benjamin T. Foster, and Matthew E. Kahn, "Improving Price Discovery to Accelerate Adaptation to Climate Change," in *Adapt or Be Adept*, 150.
51. Culp, *Structured Finance*, 30.
52. Banks, *Integrated Risk Management*, 50.
53. Krutov, *Practitioner's Perspective*, 5.
54. Krutov, *Practitioner's Perspective*, 5.

55. Richard J. Butler, "Form Regulation in Commercial Insurance," in *Deregulating Property Liability-Insurance*, 322.
56. Banks, *Integrated Risk Management*, 58.
57. Other types of captives offer alternative organizational structures, means by which multiple insureds can share ownership of a captive, and the means for those lacking the capital to form a captive to participate, but do not alter the key characteristics of a captive for our purposes.
58. Captives serve as one solution to the "cookie jar problem" that occurs if a firm simply sets aside funds to cover anticipated losses. Culp, *Structured Finance*, 525 ("The lack of credibility associated with loss reserves is often called a 'cookie jar' problem. ... Especially for firms already suffering from high costs of information asymmetries, adverse selection concerns will only exacerbate investor fears that managers may be tempted to reverse a loss reserve. And there are good reasons for worrying about this—remember the potential underinvestment problem").
59. Kathryn A. Westover, *Captives and the Management of Risk*, 3rd ed. (Dallas, TX: IRMI, 2017), 40 ("since the captive is a separate and distinct legal entity, its financial results must be reported on a standalone basis. This has two effects. The most obvious is that the cost of risks previously buried in operations results are now available for analysis by corporate. Next, the allocation of risk management costs to operations is facilitated").
60. Westover, *Management of Risk*, 41 ("The centralization of loss data using the captive as the 'loss data warehouse' in an organization provides an organization a way to track the amount, cause, and source of retained risk and the impact of risk on corporate cash flow").
61. Westover, *Management of Risk*, 42 ("By forming a subsidiary insurance company, with capital at risk, the senior management members of the organization sitting on the captive board are obliged to focus attention on the financial impact of risk on an organization. The captive is one of the best ways to present financial information about the cost of risk in ways they will understand"). Similarly, Kirzner observes that

    It can be stated with considerable confidence *that human beings tend to notice that which it is in their interest to notice.* Human beings notice 'opportunities' rather than 'situations.' They notice, that is, concatenations of events, realized or prospective, which

> offer *pure gain*. It is not the *abstract* concatenation of these events which evokes notice; it is the circumstance that these events offer the promise of pure *gain* – broadly understood to include fame, power, prestige, even the opportunity to serve a cause or to help other individuals.

Israel M. Kirzner, "The Primacy of Entrepreneurial Discovery," in Kirzner, *Ethics*, 373.

62. Westover, *Management of Risk*, 169 ("Domicile choice, however, is driven by domicile success. Risk managers don't like uncertainty of outcomes. New captive owners will go to the domiciles that have a good historical track record. They also seek a legal environment that will enable a domicile to accommodate the needs of a legitimate captive formation, where traditional loss forecasting and rating methodologies may be inapplicable").
63. Butler, *supra* note lv, at 321 ("Since many states have deregulated insurance pricing, the main source of regulatory constraint for commercial insurers is approval of changes in contract language, or form regulation").
64. *See* Cummings, "Last Bastion," 17 ("The primary normative economic rationale for policy form regulation is buyer protection. This rationale is most likely to be valid when there are significant informational asymmetries between buyers and sellers"). Such regulation is one driver of the alternative risk transfer market. Id. at 18.
65. Culp, *Structured Finance*, 538 (describing use by Eva Air of its captive as a signal of its commitment to aircraft safety and concluding that the signaling function "is not an implausible interpretation of at least some captives").
66. Andrew P. Morriss and Charlotte Ku, "Ensuring Credible Commitments," *IFC Review*, May 2022; Andrew P. Morriss and Charlotte Ku, "IFCs' 'Secret Sauce': A Commitment to an Effective Legal Infrastructure," *IFC Review*, January 2022; Charlotte Ku and Andrew P. Morriss, "IFC Regulatory Innovation: Vital to the Maintenance of a Healthy Global Financial Ecosystem," *IFC Review*, January 2022; Charlotte Ku and Andrew P. Morriss, "IFCs Act as Regulatory Capacity Builders," *IFC Review*, August 2021.
67. Edward E. Meehan, "Opportunities and Challenges Facing State Regulators Following Passage of the 1986 Risk Retention Act," *Risk Retention Reporter* 6, no. 3 (1992): 4, 5 ("the solid alternative market

facilities which a state wants to attract will be reluctant to domicile in a state whose insurance department is regarded as lax or a renegade by other regulators").

68. Thomas Holzheu and Roman Lechner, "The Global Reinsurance Market," in *Handbook of International Insurance Between Global Dynamics and Local Contingencies*, eds. J. David Cummings and Bertrand Venard (New York: Springer, 2007), 893; Robert P. Hartwig and Claire Wilkinson, "An Overview of the Alternative Risk Transfer Market," in id., 948.
69. Hartwig and Wilkinson, *supra* note lxviii, at 948.
70. Some readers may immediately worry about whether this will lead to a "race to the bottom". In general, competition among captive domiciles has primarily focused on competition to regulate appropriately rather than to simply regulate less. *See* Andrew P. Morriss, "Non-territorial special jurisdictions in the U.S. insurance market," *Journal of Special Jurisdictions* (forthcoming 2022) (describing incentives for such regulation); Tony Freyer and Andrew P. Morriss, "Creating Cayman as an Offshore Financial Center: Structure & Strategy since 1960," *Arizona State Law Journal* 45 (2013): 1297 (discussing the Cayman Islands' focus on developing regulatory infrastructure to compete for business); Ku and Morriss, *Vital* (discussing competitive pressures for quality regulation). Further, as captive specialist Karen Westover notes, "[a] captive regulatory system, ultimately, will be judged not just on its willingness to encourage captive growth but also its ability to prevent captive failures." Westover, *Management of Risk*, 182.
71. Krutov, *Practitioner's Perspective*, 7.
72. Banks, *Integrated Risk Management*, 117, n. 3. The 1989 and 1992 losses together were so severe that fifteen property and casualty insurers became insolvent as a result; this revealed a deep need for capital in the insurance industry. Id., at 124–125.
73. Niels Viggo Haueter and Geoffrey Jones, "Risk and Reinsurance," in *Managing Risk in Reinsurance: From City Fires to Global Warming*, eds. Niels Viggo Haueter and Geoffrey Jones (Oxford: Oxford University Press, 2017), 31.
74. Banks, *Integrated Risk Management*, 118.
75. Krutov, *Practitioner's Perspective*, 116 ("The importance of modeling in the analysis of insurance-linked securities is impossible to overestimate").

76. Characklis, Foster, and Kahn, "Improving Price Discovery", 159 ("The government's monitoring networks (e.g., satellites, streamflow gauges) and its financing of data collection and research provide key ingredients for private entities to process, organize, and disseminate relevant information to buyers and sellers"); Krutov, *Practitioner's Perspective*, 109 ("The annual expected loss or probability of attachment calculated by AIR Worldwide might differ, perhaps significantly, from the annual expected loss or probability of attachment if they were calculated by one of the other models based on the same data").
77. Haueter and Jones, "Risk and Reinsurance," 32 ("Catastrophe models had delivered a poor performance in New Orleans but, once they had been recalibrated, alternative investors such as hedge funds and pension funds started being more confident. This started an inflow of capital of a size that the industry had never seen before").
78. Characklis, Foster, and Kahn, "Improving Price Discovery," 160 ("Overreliance on statistical correlations based on past conditions, however, can lead to overconfidence about the strength of those relationships in the future. This is particularly problematic when we expect the future to look different from the past, as we do for many environmental financial risks in the face of climate change").
79. Jonathan Spry, "Non-Life Insurance Securitisation: Market Overview, Background and Evolution," in *The Handbook of Insurance-Linked Securities*, eds. Pauline Barrieu and Luca Albertini (Hoboken, NJ: Wiley, 2009), 14.
80. Spry, "Non-Life Insurance Securitisation," 14.
81. Banks, *Integrated Risk Management*, 129.
82. Banks, *Integrated Risk Management*, 131, 133.
83. Krutov, *Practitioner's Perspective*, 19.
84. Krutov comments that the introduction and disappearance of new products in this space reflects the difficulty of marketing complex new products when traditional insurance solutions, which are much easier to understand, are available. Krutov, *Practitioner's Perspective*, 146.
85. Krutov, *Practitioner's Perspective*, 88 ("In the case of insurance-linked securities, one of the ways to reduce the uncertainty is to improve the modelling of hurricanes and the damage they cause").
86. Krutov, *Practitioner's Perspective*, 111.
87. Characklis, Foster, and Kahn, "Improving Price Discovery", 162 ("Insurers and reinsurers have a direct, obvious financial interest in quantifying climate risk because improved information leads to

more accurate risk pricing, as well as the potential to develop new risk management products"); Krutov, *Practitioner's Perspective*, 87 ("Improvement in the ability to model catastrophe risk contributes to the development of the ILS markets. Enhanced tools give investors a higher degree of confidence and open up new options").

88. Banks, *Integrated Risk Management*, 162.
89. Andrea S. Kramer, "Critical Distinctions between Weather Derivatives and Insurance," in Culp, *Structured Finance*, 639.
90. Krutov, *Practitioner's Perspective*, 86 ("The markets remain inefficient and liquidity insufficient, but the array of options available to investors has certainly expanded").
91. Krutov, *Practitioner's Perspective*, 86 ("The ability to better model the risk has always been important in the analysis of individual securities. The better tools now available for this modelling have given investors a greater degree of confidence in the analysis and opened new options not available several years ago").
92. Krutov, *Practitioner's Perspective*, 87 (noting increased demand for modeling).
93. Krutov, *Practitioner's Perspective*, 121.
94. Krutov, *Practitioner's Perspective*, 133.
95. Krutov, *Practitioner's Perspective*, 183.
96. Krutov, *Practitioner's Perspective*, 184 ("If a month or a season is particularly warm, resulting in extra electricity demand to power air conditioning and other equipment, a CDD derivative can hedge against the economic effects of this higher-than-expected level of electricity demand. Similarly, an HDD derivative can be a hedge against the economic effects of a higher-than-expected level of demand for electricity if the weather is colder than expected").
97. Krutov, *Practitioner's Perspective*, 184.
98. Krutov, *Practitioner's Perspective*, 185.
99. Krutov, *Practitioner's Perspective*, 56 ("Catastrophe modeling has evolved in recent decades: its role is quantifying insurance risk is critical and credible. The credibility of the modelling tools continues to grow as they incorporate more and more of the latest scientific research on catastrophic events and the insurance-specific data that determines the impact of the catastrophes on insurance losses"). As Characklis, et al. note, "investors consistently describe the need for more actionable information on current climate risks, as well as the importance of reducing uncertainty over future projections. Both investors and

insurers are increasingly pushing organizations to become more transparent in evaluating and communicating their financial vulnerability to climate risks." Characklis, Foster, and Kahn, "Improving Price Discovery", 162. They also note the importance of information not just to create incentives but "because it supports decisions regarding how the risk might be most efficiently managed." Id., at 165.

100. Krutov, *Practitioner's Perspective*, 56. *See also* Tom Hutton, "The Role of Computer Modeling in Insurance Risk Securitization," in *Securitized Insurance Risk: Strategic Opportunities for Insurers and Investors*, ed. Michael Himick (Dearborn, Glenlake Publishing Company: 1998), 168 (describing increased interest in models by investors and rating agencies).
101. Characklis, Foster, and Kahn, "Improving Price Discovery," 165.
102. We have to turn to an Austrian economist because most conventional economic analyses have fallen into the same trap as "neo-Walrasian microeconomic theory" generally, they also have "virtually no scope for market entrepreneurial activity" because "in the microeconomic world peopled by market participants whose sole decision-making responsibility is that of correctly computing the solutions to constrained maximization problems, there is no scope for any speculative dimension to choice." Israel Kirzner, "Ludwig von Mises: The Man and His Economics" in *The Collected Works of Israel M Kirzner: Ludwig von Mises: The Man and His Economics*, eds. Peter J. Boettke and Frederic Sautet (Indianapolis, IN: Liberty Fund, 2019), 69. Indeed, as Kirzner elaborates, "In this approach there has been virtually no analysis of entrepreneurship because there is really no need for entrepreneurial activity. There is no need for alertness to possible future changes, because the future is, in effect, presumed already known .... Markets reach equilibrium effortlessly; no tortuous entrepreneurial process of equilibration is called for." Israel Kirzner, "Mises on Entrepreneurship," in Kirzner, *Mises*, 157.
103. Kirzner, "The Man and His Economics," 69.
104. Israel M. Kirzner, "The Subjectivism of Austrian Economics," in *The Collected Works of Israel M. Kirzner: Austrian Subjectivism and the Emergence of Entrepreneurship Theory*, eds. Peter J. Boettke and Frédéric Sautet (Indianapolis, IN: Liberty Fund, 2015), 55 ("For hitherto unnoticed opportunities to be noticed and grasped, they must be discovered; and such discovery requires a special characteristic on the part of potential discoverers, viz. the propensity to be alert"); Israel M. Kirzner,

"Roundaboutness, Opportunity, and Austrian Economics," in id., 114 ("A key insight embodied in this modern Austrian perspective concerns the role of the *hitherto unperceived opportunity*").

105. Characklis, Foster, and Kahn, "Improving Price Discovery", 172 ("In an increasingly uncertain climate world, our ability to incorporate new information quickly into market prices will be a critical determinant of adaptation success").
106. Krutov, *Practitioner's Perspective*, 84.
107. Israel M. Kirzner, "Carl Menger and the Subjectivist Tradition in Economics," in Kirzner, *Mises*, 17.
108. Kirzner, *Subjectivism*, 54.
109. Israel M. Kirzner, "Ludwig von Mises and Friedrich von Hayek: The Modern Extension of Austrian Subjectivism," in Kirzner, *Emergence*, 37 ("To *confine* ourselves and our economic analysis to the context of full mutual omniscience is not merely to accept a wildly unrealistic assumption; it is to confess that our model of the economizing world *is unable to throw light upon any process of adjustment through which, perhaps, an approach towards the fully co-ordinated state of affairs (or, indeed, of any specific state of affairs) might occur* in the real world of imperfect knowledge").
110. Israel M. Kirzner, "Uncertainty, Discovery, and Human Action: A Study of the Entrepreneurial Profile in The Misesian System" in Kirzner, *Mises*, 172.
111. Israel M. Kirzner, "Must Capitalism Yield to Socialism?" in Kirzner, *Ethics*, 34.
112. Kirzner, "Yield," 34–35.
113. Israel M. Kirzner, "Why the Market Outclasses the State," in Kirzner, *Ethics*, 38–39.
114. Kirzner, "Outclasses," 38–39.
115. Kirzner, "Outclasses," 38–39.
116. The details are set out in Scott E. Harrington, Greg Niehaus, and Kenneth J. Risko, "Enterprise Risk Management: The Case of United Grain Growers," in Culp, *Structured Finance*, 745.
117. Richard Rudolph, *A Tedious Brief History of Insurance* (Austin, TX: National Alliance Research Academy, 2015), 205.
118. Krutov, *Practitioner's Perspective*, 450.
119. Characklis, Foster, and Kahn, "Improving Price Discovery", 174.
120. Characklis, Foster, and Kahn, "Improving Price Discovery", 174.
121. Kirzner, "Primacy," 367.

122. Israel M. Kirzner, "The Theory of Entrepreneurship in Economic Growth, " in Kirzner, *Essence*, 305.
123. Israel M. Kirzner, "The Perils of Regulation: A Market-Process," in Kirzner, *Ethics*, 435–436.
124. Kirzner, "Perils," 436.
125. See, e.g., Terry L. Anderson, "Donning Coase-Coloured Glasses: A Property Rights View of Natural Resource Economics," *Australian Journal of Agricultural & Resource Economics* 48 (2004): 445. *See also* Andrew P. Morriss, "The Opthamologist is In: Prescribing 'Coase-Colored Glasses,'" in *Save the Environment—Unleash the Markets! Essays in Honor of Terry L. Anderson*, ed. Kurt R. Leube (Jena, Germany: TvR Medienverlag, 2021) (surveying Anderson's contributions).
126. Terry L. Anderson, "Viewing Land Conservation Through Coase-Colored Glasses," *Natural Resources Journal* 44 (2004): 125, 134.
127. Krutov, *Practitioner's Perspective*, 5 ("In describing insurance markets, it is also important to note that insurance is one of the most heavily regulated industries, a fact that, by itself, introduces a broad set of constraints and risks not found in other sectors"); Martin F. Grace and Hal S. Scott, "An Optional Federal Charter for Insurance: Rationale and Design," in *The Future of Insurance Regulation in the United States*, eds. Martin F. Grace and Robert W. Klein (Washington, DC: The Brookings Institution, 2009), 66 ("If products are approved quickly, then firms can compete more efficiently on product innovation and design. However, if products are approved slowly, the incentive for insurers to develop and market new ideas is reduced").
128. Lawrence G. Brandon, *Sound a Clear Call: Unveiling the Future of the Insurance Industry* (Malvern, PA: CPCU-Harry J. Loman Foundation, 1984), 92.
129. Brandon, *Clear Call*, 93.
130. Characklis, Foster, and Kahn, "Improving Price Discovery", 174–175.
131. Banks, *Integrated Risk Management*, 47.
132. Banks, *Integrated Risk Management*, 59–60.
133. Banks, *Integrated Risk Management*, 57.
134. Culp, *Structured Finance*, 22.
135. Culp, *Structured Finance*, xv.
136. *See* Morriss, "Non-territorial."
137. *See* Morriss, "Non-territorial."
138. Karen Cutts, "Purchasing Group Offers Insureds Best of Both Worlds," *Risk Retention Reporter*, 2, no. 8 (1988), 5.

139. 15 U.S.C. §3901 et seq.
140. 15 U.S.C. §8201 et seq.
141. Public Law 97–45, 95 Stat. 949, §2(a)(4) (1981) (allowing Bermuda and Cayman Islands domiciled companies to qualify).
142. *See* Erin A. O'Hara and Larry E. Ribstein, *The Law Market* (Oxford: Oxford University Press 2009).
143. Kramer, "Critical Distinctions," 639.
144. Krutov, *Practitioner's Perspective*, 87–88 ("At this point, however, most investors do not utilize the tools already available, and many make their investment decisions based primarily on judgment and a back-of-the-envelope type of analysis. While there are some extremely sophisticated players in this market, there is significant room for improvement in investor understanding and modelling of catastrophe insurance-linked securities").
145. Israel M. Kirzner, "Entrepreneurship and American Competitiveness," in Kirzner, *Ethics*, 516.
146. Characklis, Foster, and Kahn, "Improving Price Discovery", 174–175.

A Classical Liberal Case for Target-Consistent Carbon Pricing

Ed Dolan

Property rights and prices serve as the foundational principles of classical liberal economics. Property rights draw the boundaries of what parties with disparate interests in the use of scarce resources may and may not do. The price system provides a key mechanism for sharing information about the values that owners and nonowners place on those resources. But how to translate those principles into a workable mechanism for dealing with the notoriously wicked issue of climate change, where property rights are not clearly established, prices do not emerge spontaneously from a coherent market process, and the whole problem is embedded in a web of scientific and technological uncertainties?

E. Dolan (✉)
Niskanen Center, Washington, DC, USA
e-mail: edolan@niskanencenter.org

J. H. Adler (ed.), *Climate Liberalism*, Palgrave Studies in Classical Liberalism,
https://doi.org/10.1007/978-3-031-21108-9_12

Property Rights and Their Limits[1]

There is no better place to start than with this famous passage from John Locke's Second Treatise of Government[2]:

> Though the earth, and all inferior creatures, be common to all men, yet every man has a property in his own person: this nobody has any right to but himself. The labour of his body, and the work of his hands, we may say, are properly his. Whatsoever then he removes out of the state that nature hath provided, and left it in, he hath mixed his labour with, and joined to it something that is his own, and thereby makes it his property. It being by him removed from the common state nature hath placed it in, it hath by this labour something annexed to it, that excludes the common right of other men: for this labour being the unquestionable property of the labourer, no man but he can have a right to what that is once joined to, at least where there is enough, and as good, left in common for others.

A few pages later, Locke makes it clear that the principle for acquiring property applies not just to acorns gathered in the woods, wild game, or water from a stream, but to the land itself[3]:

> As much land as a man tills, plants, improves, cultivates, and can use the product of, so much is his property. He by his labour does, as it were, inclose it from the common.

From these passages we can distill three rights and three corresponding duties:

Rights:

- to property in one's own person
- to property in the fruits of one's own labor
- to property in land and natural resources taken from the commons by first use

Duties:

- to abstain from harming others in their persons
- to abstain from taking or harming the property of others
- to leave enough and as good for others when taking from the commons

The rights and duties are inseparable. To claim the former, one is obliged to uphold the latter.

Locke's third right, that of acquiring property from the commons by mixing one's labor with it, has come to be known variously as enclosure, homesteading or, most broadly, acquisition by first use. Over the years, acquisition by first use has been extended to more abstract forms of property, such as water rights or rights to dispose of wastes in the earth's atmosphere.

There is, however, an important issue to be dealt with here. In discussing water from a stream, Locke clearly had in mind a situation in which the amount withdrawn was trivial compared to the flow of the stream as a whole, so that the stream would be constantly replenished. In that case, the right would be to draw water at will at a rate consistent with replenishment. On the other hand, if it would be a matter of withdrawing a large amount of water in a more arid area, the claim by first use might have to be limited to a specific quantity, so that there would be enough flow left for others to stake a claim. As discussed further below, a point might be reached where the possibility not only of new claims but of the continued exercise of previously legitimate claims, would be exhausted.

Carbon emissions, which persist for centuries, and are impracticable to reverse at large scale with any technologies available today, more closely fit the case where replenishment is impossible. By now, we have passed the point where the waste disposal capacity of the atmosphere has been effectively used up. As a result, emissions that are now occurring are in violation not only of the Lockean duty to leave enough and as good for others, but also of the duties not to harm others in their persons and property.

Historically, some air pollution may have been treated like the first case, so that an industrialist who built a steel mill in a sparsely populated region would homestead the right to emit wastes from the facility indefinitely at the rate that was originally established. However, that model does not fit carbon emissions, which persist for centuries, and are impracticable to reverse at a large scale with any technologies available today. It is closer to the truth to say that from the very start, each bit of carbon emitted has used up a part of the atmosphere's waste disposal capacity.

Today, then, the question of whether some early first use claims to carbon emissions might have been justifiable is of only theoretical interest. Sometime between Newcomen's first practical steam engine and the present we have passed the point where the waste disposal capacity of the atmosphere was effectively used up. Emissions that are now occurring are in violation not only of the Lockean duty to leave enough and as good for others but also of the duty not to harm others in their persons and property.

What would Locke have had us do in such a situation?

The Role of Government

As it turns out, Locke had to deal with this same question in his own day, not as applied to carbon emissions but to land. Here is his answer[4]:

> It is true, in land that is common in England, or any other country, where there is plenty of people under government, who have money and commerce, no one can inclose or appropriate any part, without the consent of all his fellow-commoners; because this is left common by compact, i.e. by the law of the land, which is not to be violated. And though it be common, in respect of some men, it is not so to all mankind; but is the joint property of this country, or this parish. Besides, the remainder, after such enclosure, would not be as good to the rest of the commoners, as the whole was when they could all make use of the whole.

When Locke refers to "the consent of all fellow-commoners," he does not mean unanimous consent by some committee of the whole. Instead, he has in mind consent as exercised through the institutions of government. Not just any kind of government, but a government that is limited in three important ways.

First, this government's principal purpose—in fact, its *only* purpose—would be the protection of property: "Government has no other end but the preservation of property."[5] "The great and chief end, therefore, of men's uniting into commonwealths, and putting themselves under government, is the preservation of their property."[6] Importantly, under the principle of self-ownership, "property" for Locke included not just the estates and manors of the landed rich, but the personal security and livelihoods of everyone. "Every man has a property in his own person," etc.[7]

Second, Locke described what would later come to be called a *liberal* government. Such a government would be characterized by the separation of executive and legislative powers. It would also follow the rule of law, that is, of "settled standing rules, indifferent, and the same to all parties."[8] It would also have an independent judiciary. To ensure just enforcement of the law, no one should be the judge in their own case.

Third, recognizing that it would be impracticable to secure the actual consent of every citizen for every decision, Locke understood that the ordinary business of government should be conducted by majority rule. Even so, the actions of the government should command at least the tacit support of all citizens.[9] By tacit consent, Locke meant that anyone who accepts the benefits of living under a government, whether those benefits take the form of protection of landed property or simply the use of public highways, must assent to be governed by all the laws of that government without picking and choosing.

Locke's justification for the establishment of such a government was largely pragmatic. He recognized that government involved a tradeoff between the freedoms of the state of nature and the benefits of a more settled way of living:

> For being now in a new state, wherein he is to enjoy many conveniencies, from the labour, assistance, and society of others in the same community, as well as protection from its whole strength; he is to part also with as

> much of his natural liberty, in providing for himself, as the good, prosperity, and safety of the society shall require; which is not only necessary, but just, since the other members of the society do the like.[10]

Such, then, are Locke's views regarding property and government. As applied to the issues we face today, they provide a sound, classical liberal justification for action to protect people and their property from harm caused by carbon pollution. But we need more than just a philosophical justification. We also need practical procedures for carrying out the government's obligations. For that, we must turn to later classical liberal writers.

Friedrich Hayek's Contributions[11]

Although Friedrich Hayek never wrote a word about climate policy, his work is highly pertinent to the debate over carbon pricing. This section will focus on three key points that he made regarding the role of science in public policy, the role of the price system as a carrier of information, and the role of regulation when markets fail to do their job.

In a famous 1945 article, Hayek distinguished between two types of knowledge.[12] The first, "knowledge of the particular circumstances of time and place," is, by its nature, widely distributed among individuals, each of whom sees only a small part of the whole picture. The second, scientific knowledge, is less widely dispersed. As Hayek put it,

> as far as scientific knowledge is concerned, a body of suitably chosen experts may be in the best position to command all the best knowledge available— although this is of course merely shifting the difficulty to the problem of selecting the experts.

In a later essay, Hayek pointed to respect for scientific knowledge, represented by a consensus of reputable experts, as a key trait that distinguished classical liberal thinking from that of conservatives.[13] Liberals, said Hayek, are prepared to come to terms with new scientific knowledge, whether they like its immediate effects or not. Conservatives, in contrast, are warier of science:

> Personally, I find that the most objectionable feature of the conservative attitude is its propensity to reject well-substantiated new knowledge because it dislikes some of the consequences which seem to follow from it—or, to put it bluntly, its obscurantism. I will not deny that scientists as much as others are given to fads and fashions and that we have much reason to be cautious in accepting the conclusions that they draw from their latest theories. But the reasons for our reluctance must themselves be rational and must be kept separate from our regret that the new theories upset our cherished beliefs.

But neither scientific knowledge nor the knowledge of time and place is enough alone. There has to be some way to bring them both to bear—some way for local actors to obtain the further information about global scarcities that they need to "fit [their] decisions into the whole pattern of changes of the larger economic system."[14] In Hayek's view, the price system is the mechanism that fills that function. If the price of an input you use for your farm or factory goes up, you don't need to know why—you only need to know that it has somehow become scarcer, and that you should, if possible, use less of it. Similarly, if the price of something you produce goes up, all you need to know is that you should do your best to produce more.

But when we apply that reasoning to the problem of climate, we encounter a problem. There is a critical scarce resource—the ability of the earth's atmosphere to absorb carbon emissions—for which the market does not spontaneously generate a price.

Hayek was well aware of the problem of missing prices. In yet another work, *The Road to Serfdom,* he wrote,

> Where, for example, it is impracticable to make the enjoyment of certain services dependent on the payment of a price, competition will not produce the services; and the price system becomes similarly ineffective when the damage caused to others by certain uses of property cannot be effectively charged to the owner of that property ... In such instances

> we must find some substitute for the regulation by the price mechanism. But the fact that we have to resort to the substitution of direct regulation by authority where the conditions for the proper working of competition cannot be created, does not prove that we should suppress competition where it can be made to function.[15]

Hayek's invocation of "regulation by authority" is very much in the spirit of Locke, who, as we have seen, assigned to the government the role of managing those resources that remain in common once the limits of appropriation by first use have been reached. But Hayek and Locke both leave unanswered the question of exactly what kind of laws or regulations are appropriate in the case of climate policy. We turn to that in the next section.

Pricing, Science, and the Future[16]

Regulation by authority could mean a lot of things. It could, for example, mean central planning in which a government body issues specific quantitative and technological directives to producers and consumers. Such a body could insist that all facilities that emitted CO_2 in excess of some minimal threshold install carbon capture and storage technology. It could impose fuel efficiency standards on all vehicles, and mandate that a gradually rising percentage of vehicles use zero-emission technologies. It could restrict the use of natural gas for heating and cooking in newly built homes and apartments, and so on.

However, it is likely that Hayek would have regarded such measures skeptically. He would have objected that a government agency, however strong its staff of scientific experts, would not have access to the knowledge of time and place that are necessary to fine-tune restrictions on emission to widely varying circumstances. Even if they had access to such information, Hayek would have doubted their ability to process it. The whole Austrian tradition in economics out of which Hayek came spent much of the first half of the twentieth century in a long-running debate with advocates of socialist planning. In short, he would have viewed direct command-and-control regulation as clumsy and inefficient, suitable for use only as a last resort, and to be avoided where there was any way that competition could be made to function.

Fortunately, there is indeed a way to bring prices to bear on the problem of carbon pollution. That alternative is to use the powers of government to set a price on emissions of CO_2 and equivalent. Decentralized decision makers, from power plants to individual drivers, would then tailor their actions to the combined influence of the centrally generated price signal and their own local knowledge. Although not perfect, price-based policies offer significant practical advantages over central planning while more faithfully honoring the philosophical tradition of classical liberalism. Modern classical liberals have sometimes seen carbon prices as "user fees" levied on emitters for use of the atmospheric commons that are the properties of all mankind, although I am unaware that Hayek ever made that argument.

Just how would such a policy work in practice? The approach favored by many economists begins by estimating something called the *social cost of carbon* (SCC). The SCC is a measure of the total harm, now and in the future, caused by the emission of one ton of CO_2 or the equivalent of other greenhouse gasses. A tax per ton of emissions can then be imposed at a rate equal to the SCC. Sweden, Denmark, and the Canadian province of British Columbia are among the many countries and jurisdictions that have imposed such carbon taxes.

A variant is to generate a carbon price equal to the SCC by issuing a fixed number of emission permits, each allowing a specific quantity of emissions. The emissions are then traded among pollution sources. Those with low abatement costs become sellers, those with high abatement costs become buyers, and the price at which permits are traded becomes the carbon price. The number of permits is adjusted as necessary to keep the price of permits close to the SCC. That approach is known as *emissions trading* or *cap-and-trade*. The European Union, China, and the U.S. state of California are among the many examples of jurisdictions that have implemented emissions trading.

Whether taxation or trading is chosen, however, this "prices-first" approach to carbon pricing requires some way of calculating the SCC. The standard method relies on elaborate *integrated assessment models*

(IAMs). These are systems of equations that span multiple disciplines, including economics, demographics, climatology, energy science, and more. The IAMs incorporate assumptions regarding the sensitivity of global temperature to atmospheric CO_2 concentrations, the effects of temperature on rainfall and destructive storms, as well as socioeconomic assumptions such as savings rates and population growth.[17]

There is a problem with the prices-first approach, however. The use of IAMs to determine a carbon price sounds very objective and scientific, but on closer inspection, it turns out that the models leave great room for subjectivity. One critical point concerns the value that should be put on harms and benefits that occur in the distant future compared to those that occur today.

The parameter in the IAM that allows comparisons of values at different points in time is called the *discount rate.* For example, using a discount rate of 3% would mean that the value of $100 in costs or benefits incurred one year from now would be equal to $97 incurred today. As the prospective cost or benefit recedes farther into the future, its present value decreases. Over the long time horizons that are relevant to climate change, the decrease is dramatic. Discounted at 3%, the present value of $100 one hundred years from now is only about five cents.

In practice, the social cost of carbon is more sensitive to the discount rate than to any other single parameter of the IAM. For example, during the Obama administration, an interagency working group, using a discount rate of 3%, estimated the SCC to be $51 per ton of CO_2, a number that would imply a relatively aggressive carbon tax. But just a few years later, the Trump administration lowered the estimate to just $1 per ton.[18] Such a value would justify essentially no action to address climate change. Although some other elements of the IAM were also tweaked, the biggest difference between the Obama and the Trump estimates was that the Trump team raised the discount rate to 7%. Discounted at 7% rather than 3%, the value of $100 a hundred years out is not 5 cents but just one-tenth of one cent.

Really, the choice of a discount rate is not a "scientific" choice at all. It has no unique value that can be discovered, or even approximated, by means of reason and evidence. Instead, the reality is that some people care a lot about the welfare of generations and are prepared to make sacrifices on their behalf. Other people care just as much about future generations but think that they will be richer, will have access to better technology, and will be able to take care of themselves. Still, others just don't care much about what happens in the distant future. You can translate these views into the language of higher or lower discount rates, but whether you express them quantitatively or qualitatively, they still fall more in the category of "cherished beliefs" than calculations that we can comfortably assign to a body of experts.

The teams that construct IAMs deal with this problem in various ways. Some base their discount parameters on market interest rates, even though there are no market rates that extend 100 years into the future, and shorter-term rates vary so widely that choosing among them is itself a highly subjective exercise. Some IAMs include assumptions like infinitely lived individuals, or overlapping generations, each of which cares as much about the next generation as about their own. But in end, these are only ways of hiding the variety and subjectivity of attitudes toward the future behind a veil of numbers.

What is more, the discount rate is not the only point at which IAMs must deal with subjectivity. For example, in addition to more or less objective damages from crop loss or storm damage, some environmentalists feel strongly that the SCC should reflect more subjective harms like the loss of biodiversity. Others think harm to the homes or livelihoods of poor farmers and herders should be weighted more highly, dollar for dollar, than harm to residents of wealthy countries—especially since emissions from those countries have caused most of the global warming to date. They advocate some kind of "equity weighting" to deal with this issue. Still, others think extra weight should be given to low-probability but disastrous events that might end civilization as we know it. That is sometimes known as the "fat-tail problem," since those risks lie far out in the tails of the probability distributions that are constituent parts of the IAMs. It is, however, difficult to measure and price these risks through central calculation, and markets are not very good at doing the job either.[19] There are reasonable arguments in favor of taking all of these considerations into account, but doing so only moves the process of estimating the SCC farther away from scientific rationality and more toward subjectivity.

The Case for Target-Consistent Pricing

But there is an alternative to the prices-first approach—one that begins from the social cost of carbon. That alternative, known as the *target-consistent* approach, provides a better way of dealing with the inherent subjectivity of climate policy.

A target-consistent policy begins by defining the goal that policymakers would like to accomplish, without making any claim that the chosen target is optimal or fully objective, and then works backward to develop an appropriate carbon price to reach the goal.

In practice, world leaders have already embraced the target-consistent approach. The Paris Agreement, which 192 countries have signed on to, is an example. The goal of that agreement is "to limit global warming to well below 2, preferably to 1.5 degrees Celsius, compared to pre-industrial levels." There is a consensus that to reach that target, carbon emissions will need to fall to net-zero by about 2050. Participants in the Paris agreement by no means ignored what science had to say, but the actual target they agreed on was openly represented as a political compromise.

Once the target is set, technical experts can get to work setting an appropriate carbon price that would put emissions on track to reach it. The price could be implemented either as a carbon tax or through emissions trading. The target-consistent carbon price could, if desired, be supplemented by administrative measures such as performance standards, accelerated permitting for energy transmission, research subsidies, and more. Those who make a case for a mixed strategy point out that a strong price signal can increase the effectiveness of performance standards or other regulatory measures. In other cases, policies such as energy standards for home appliances can overcome behavioral issues that reduce consumer sensitivity to pricing.[20]

Target-consistent carbon pricing, sure, also has its critics. To some, it presents yet another variant of the knowledge problem. Others worry that it would produce dynamic responses that are impossible to predict. They also wonder how it would handle technological breakthroughs that could help but that would also be impossible to predict. These are valid concerns. Advocates of target-consistent pricing have taken care to address them in detail.

For example, if the target-consistent approach were implemented as a carbon tax, Noah Kaufman and several colleagues advocate an approach they call "adaptive management."[21] Based on the chosen target, the Kaufman approach would calculate an initial time-path for the carbon price using an IAM. However, that price-path is only an initial working approximation. As unforeseeable technological changes and dynamic responses developed, the price-path would periodically be adjusted to keep the system on track to the chosen target.

Instead, if the emissions-trading approach to target-consistent pricing were chosen, the first step would be to employ a team of climate scientists to translate the chosen target into a carbon budget. For example, the carbon budget might specify that the United States' share of allowable future emissions would be 75 gigatons of CO_2. That in turn could be broken down into, say, 75 million emission permits each allowing one kiloton of emissions. The permits could be placed on the market by auction, grandfathering, or any other democratically chosen method. The market would then spontaneously generate the needed carbon price. Buyers would include parties with urgent needs to continue emissions while cleaner technologies were developed and speculators who guessed that certificate prices would rise over time. Conservation organizations might also buy permits to hold them unused in order to reach the net-zero target at an earlier date. Sellers might include industries that decided that selling their permits would generate enough revenue to pay for going green immediately. They might also include speculators who were optimistic about the pace of technology and therefore expected permit prices to fall over time.[22]

On a more fundamental level, some critics have objected that the target-consistent approach "replaces scientific assessments ... with subjective judgments about policy targets and choices."[23] However, the above discussion shows the charge to be misplaced, both for the tax and the trading variants. The processes used to set the target itself and the carbon price consistent with it would take all available scientific and economic knowledge into account. The difference from a prices-first,

SCC-based policy is that the target-consistent variant would be transparent about the elements of the policymaking process that are inherently subjective, rather than burying them in equations representing discount rates, equity weights, risk weights and as is done in an IAM.

Target-consistent pricing can stand on its technical merits alone, but it seems to me that it has special appeal to those of a classical liberal mindset. The central role of prices in achieving the chosen target creates incentives across every margin to reduce emissions, while simultaneously foreclosing few possibilities as administrative regulations sometimes do. Both variants of target-consistent pricing incentivize innovation, and both contain flexible mechanisms for adapting to innovation as it emerges. The emissions-trading variant has the added philosophical elegance of allowing the carbon price to emerge spontaneously from market interactions without the need for central price calculations of even the flexible, adaptive type.

Conclusions

The task set at the outset of this chapter was to outline an approach to climate policy that is consistent with the classical liberal tradition. Here is a recap:

First, expressed in Lockean terms, the problem of climate policy is one of managing the unenclosed atmospheric commons that remain after reaching or exceeding the point where further carbon emissions are impossible without causing harm to people and property.

Second, again following Locke's reasoning, the responsibility for managing the commons properly lies with all of the commoners acting through the agency of a government that operates under the principles of liberal democracy.

Third, for reasons elaborated by Hayek, climate policy cannot simply be delegated to a committee of experts. Effective action against climate change requires the integration of expert scientific knowledge with decentralized knowledge of time and place. That can best be accomplished through the price system. However, since no price on carbon emissions emerges spontaneously in a market economy, establishing a carbon price requires action by the government.

Fourth, climate policy should be informed by science to the greatest extent possible, but science alone cannot provide all the answers. Policymakers must also make subjective judgments on issues that include discounting, risk weighting, equity weighting, and more. That makes carbon pricing an inherently political task. The same goes for any supplementary administrative or regulatory measures if a mixed strategy is chosen.

Putting all this together, the conclusion is that neither government action nor the spontaneous operation of free markets alone is sufficient for an effective response to the climate challenge. Both must work together. Climate policy is an inherently political task. The best that a classical liberal can hope for is that the politics are those of a liberal government. There can be no pretense that the outcome will be optimal. No matter what package of policies is chosen, it will be too weak to satisfy some and too stringent to satisfy others. But doing something will be better than doing nothing.

Notes

1. The material in this section is based in part on Ed Dolan, "Taking Locke Seriously: Of Government, Property Rights, and Climate Change," Niskanen Center, Dec. 10, 2020, https://www.niskanencenter.org/taking-locke-seriously-of-government-property-rights-and-climate-change/.
2. John Locke, 2010 (1690), Second Treatise of Government, Project Gutenberg EBook (Ch. 5, Sec 27). https://www.gutenberg.org/files/7370/7370-h/7370-h.htm. Chapter and section references are consistent with the 1690 edition.
3. Locke, Second Treatise of Government, (Ch. 5, Sec. 32).
4. Locke, Second Treatise of Government, (Ch. 5, Sec. 35).
5. Locke, Second Treatise of Government, (Ch. 7, Sec. 34).
6. Locke, Second Treatise of Government, (Ch. 9, Sec. 124).
7. Locke, Second Treatise of Government, (Ch. 5, Sec. 7).
8. Locke, Second Treatise of Government, (Ch. 7, Sec. 87).
9. Locke, Second Treatise of Government, (Ch. 8, Sec. 199–120).

10. Locke, Second Treatise of Government, (Ch. 9, Sec. 130).
11. The material in this section is based in part on Ed Dolan, "Friedrich Hayek on Carbon Taxes," Niskanen Center, April 12, 2017, https://www.niskanencenter.org/friedrich-hayek-carbon-taxes/.
12. Friedrich Hayek, "The Use of Knowledge in Society," *American Economic Review* 35, no. 4 (September, 1945): 519–530, Online version published by Econlib, Feb. 5 2018, https://www.econlib.org/library/Essays/hykKnw.html.
13. Friedrich Hayek, "Why I am Not a Conservative," in *The Constitution of Liberty* (Chicago: University of Chicago Press, 1960), https://www.cato.org/sites/cato.org/files/articles/hayek-why-i-am-not-conservative.pdf.
14. Hayek, Use of Knowledge.
15. Friedrich Hayek, *The Road to Serfdom* (Chicago: University of Chicago Press, 1944), p. 40. Full text available online at https://archive.org/details/in.ernet.dli.2015.46585.
16. The material in this section and the next is based in part on Ed Dolan, "Are Economists Wrong about Carbon Prices," Niskanen Center, March 24, 2022, https://www.niskanencenter.org/are-economists-wrong-about-carbon-pricing/.
17. For a more detailed overview of IAMs, see "Q&A: How 'integrated assessment models' are used to study climate change," Carbon Brief, October 2, 2018. https://www.carbonbrief.org/qa-how-integrated-assessment-models-are-used-to-study-climate-change/.
18. "How the Trump Administration Is Obscuring the Costs of Climate Change," Institute for Policy Integrity, March 2018. https://policyintegrity.org/files/publications/Obscuring_Costs_of_Climage_Change_Issue_Brief.pdf.
19. See, for example, Gernot Wagner et al., "Eight Priorities for Calculating the Social Cost of Carbon," *Nature* 590 (February 19, 2021): 58–50, https://www.nature.com/articles/d41586-021-00441-0.
20. For additional examples of how pricing and regulatory measures can work together, see Ed Dolan, "The Underappreciated Versatility of Carbon Pricing," Niskanen Center, October 4, 2021, https://www.niskanencenter.org/the-underappreciated-versatility-of-carbon-pricing/.
21. Noah Kaufman et al., "A Near-Term to Net Zero Alternative to the Social Cost of Carbon for Setting Carbon Prices," *Nature Climate Change* 10, no. 11 (2020): 1010–1014, https://www.nature.com/articles/s41558-020-0880-3.

22. For details of target-consistent emissions trading, see Ed Dolan, "The Role of Carbon Pricing in Deep Decarbonization," Niskanen Center, August 26, 2021, https://www.niskanencenter.org/the-role-of-carbon-pricing-in-deep-decarbonization/.
23. Joseph E. Aldy et al., "Keep Climate Policy Focused on the Social Cost of Carbon," *Science* 373, no. 6557 (August 19, 2021): 850–852, https://www.science.org/doi/10.1126/science.abi7813.

Climate Change, Political Economy, and the Problem of Comparative Institutions Analysis

Mark Pennington

Introduction

Within the field of political economy, it is now widely recognised that judgements about the 'success', or 'failure' of institutions should be comparative and empirical in nature. Institutions should not be considered to 'fail' in relation to some idealised alternative state but must be judged in comparison to the actualised performance of a set of 'real-world' alternatives. In the case of managing anthropogenic climate change, however, this chapter will suggest that the necessary empirical comparisons between real-world institutional alternatives *cannot* easily be generated. The analysis suggests that the character of the climate governance problem blocks the production of the comparisons between actualised, as opposed to idealised governance alternatives that the comparative institutional approach recommends.

M. Pennington (✉)
Department of Political Economy, King's College London, London, UK
e-mail: mark.pennington@kcl.ac.uk

J. H. Adler (ed.), *Climate Liberalism*, Palgrave Studies in Classical Liberalism,
https://doi.org/10.1007/978-3-031-21108-9_13

The discussion is divided into three sections. The first sets out the case for comparative institutions analysis in political economy and the role of empirical comparisons in such analysis. The focus here will be on explaining why 'ideal' and 'asymmetrical' forms of institutional reasoning should be rejected in favour of a comparative approach that examines the empirical differences between 'real world' institutional alternatives and their capacity to address problems of incentive compatibility and the cognitive limitations of decision-makers. Section "Climate Change Governance and Comparative Institutions Analysis" sets out some of these problems in the context of the alternative institutional responses that have been proposed to address the climate governance dilemma. Section "The Problem of Comparative Evaluation" proceeds to show that the background conditions required for a robust account of the costs and benefits associated with these institutional alternatives are largely absent.

Comparative Institutions Analysis in Political Economy

Normative analyses in political economy have increasingly come to recognise the importance of comparative institutions reasoning. If people are perfect or perfectible then there would be no need for institutional analysis of any kind because 'good outcomes' would be generated *irrespective* of the institutional setting. It is however precisely because of human imperfections or 'non-ideal' conditions that institutions matter. Given the constraints presented by human imperfections some institutions may generate better outcomes than others, and it is the task of political economy analysis to discern what those institutions might be.

The development of the comparative institution perspective arose largely in response to the failure of post-war neoclassical economics to address adequately *how* institutions affect the content of decision-making. Though neoclassical theory *appears* to involve an evaluation of institutional regimes it does so in a manner that ignores the institutional qualities at issue. This tendency was and still is equally apparent in schools of thought enthusiastic about market economies and in those advocating a larger role for the state. In the former instance, the Chicago school tradition works with assumptions of perfectly informed actors and perfectly aligned incentives to argue that market institutions produce optimal results—with any departures from optimality reflecting real costs not worth the price of correcting. No attempt is made by these economists to explain how market actors acquire the capacity to coordinate their actions in this manner and why governmental actors may not—the assumptions of the full information general equilibrium model are simply equated with the situation of market participants.[1] Critics of Chicago policy conclusions are, however, guilty of the very same tendency. The assumption of perfectly informed and motivated governmental actors leads to the view that state intervention produces optimal results when incentives in markets are misaligned. While recognising that real-world markets are sub-optimal no explanation is given as to why governmental actors can bring about a more optimal set of decisions—it is simply assumed that they can.

In response to this state- of- affairs, contributions from the comparative transaction costs approach of Coase[2] and Demsetz,[3] public choice theory following Buchanan and Tullock,[4] the Bloomington school in the tradition of Elinor and Vincent Ostrom,[5] and the Austrian school of economics following Hayek,[6] have developed an alternative toolkit for comparative institutions reasoning. This toolkit deploys a symmetrical account of the actors operating in different institutional arenas and an explanation of how the institutions in question may work *given the imperfections of the relevant agents.* Thus, if actors in markets are assumed to be self-interested and potentially opportunistic it should not be assumed that those in the public sector are immune from self-interested concerns and the temptation to abuse power, without an explanation of what it is about public sector environments that better aligns incentives or that leads to some transformation of human motivations. Similarly, if actors in markets are assumed to be imperfectly informed it should not be assumed that those in the public sector are better informed without an explanation of what it is about public sector institutions that may or may not enable people working within them to access more information than market participants. Second, and in tandem with this focus on symmetry of assumptions judgements of institutional 'failure' or 'success', must contain an empirical component. The performance of markets, public sector organisations or other institutional configurations should not be judged against a theoretical ideal standard but against the *actualised* performance of 'real world' alternatives.[7]

Drawing on such comparative institutions reasoning writers in the Coasian and public choice traditions have argued that systems replacing private property rights with large-scale public ownership may often *exacerbate* information asymmetries, principal-agent problems and various externality or collective action dilemmas that are routinely found in markets. Citizens have few incentives to be well informed about the effects of many public decisions because in a large number electorate the chance that their decision to become better informed will affect who wins an election is infinitesimally small.[8] In more market-oriented settings by contrast, the 'exit' option allows actors the space to contract around the externality producing actions of others even if they cannot avoid them entirely. The capacity to minimise exposure to externalities also incentivises agents to be better informed about their commercial and jurisdictional choices since these will have more influence on personal outcomes than in a collective choice setting with no exit option. Such analyses while offering a theoretical explanation for the comparative advantages of market systems have in turn been substantiated by empirical observations of the chronic institutionalised corruption and 'rent seeking' that has accompanied large-scale 'real world', attempts to replace private with public property.

Where Coaseans and public choice theorists focus on comparative incentives the Austrian tradition has analysed problems of bounded rationality arising from the constitutional limits of human intelligence.[9] On this view, in anything beyond the simplest of societies even the most intelligent and well-informed actors will be relatively ignorant of the society of which they are a part, and this means that the likelihood of error is high. Error here is not simply a failure to gather information about preferences, resource scarcities and organisational models, which is known in principle to be available—a failure of information search—but reflects a problem of radical ignorance where agents 'do not know what they do not know'. Though market and governmental actors suffer from this radical ignorance for Austrians competitive and decentralised forecasting is likely to draw upon a wider pool of knowledge than centralised forecasting by state monopolies and though it cannot eliminate error the dispersal of decision-making authority in markets reduces the likelihood of any single forecasting error affecting the entire society. Beyond their inability to predict the future it is unlikely that central planners can spot and respond to as many potential gaps or opportunities in the economic environment as effectively as a multitude of dispersed property owners with the freedom to exit from and enter into, different commercial transactions. It is the profits and losses from such transactions that continually signal which paths should be pursued and which should be avoided. As with the Coasian and public choice contributions, while theoretical in character this Austrian analysis is also grounded in empirical observations of the chronic shortages and surpluses that have plagued 'really existing' planned economies and the tendency for more liberal economic regimes to exhibit higher levels of economic dynamism.

Though there appears to be a strong comparative institutions case for market economies over centrally managed alternatives, writers in the Bloomington tradition have shown that for a variety of common pool and collective good problems the operation of private markets and systems of public property alike may be outperformed by institutions that exist outside the dichotomy between markets and states. These include various forms of localised communal control and regulation or mixed property regimes that develop rules for the management of resources from the 'bottom up'. These governance models typically arise in a context of polycentric structures, where competing and overlapping decision centres experiment with different mixes of private, communal, and public property and where no single model of governance can be imposed from the 'top down'. 'Bottom up' rule formation is more likely to align incentives in ways that reduce the presence of external costs because relative to a monocentric or top-down system those designing the rules will be in closer proximity to those subject to them. In addition, the cognitively limited and fallible actors experimenting with any given set of rules are limited in their capacity to impose possible mistakes on actors outside of their jurisdiction. Again, though theoretical in nature these comparative institutional speculations have been substantiated through the observation of the different outcomes generated under more, or less polycentric arrangements.[10]

These empirical differences in performance between planned economies and markets and between polycentric and monocentric governance regimes appear relatively clear but it is important to recognise that the capacity to arrive at such comparative evaluations and the potential to revise them, has been dependent on certain 'meta-level' conditions that have allowed the production of the relevant comparative institutional 'data'. One factor that has enabled this data production has been the possibility for parallel or simultaneous comparisons between institutions facilitated by a global context where different modes of governance *coexist* and where observed counterfactuals across time and space may be generated. This context equates to a 'meta-level' form of polycentricity that mimics to some extent the experimental conditions found in the natural sciences where observed differences in outcomes between relatively similar societies such as the former East and West Germany, contemporary North and South Korea, and contemporary Venezuela and Chile, might reasonably be attributed to institutional differences. A further advantage of this meta-level polycentricity is that it allows for an *ongoing* process of data production, institutional reflection, and evaluation to unfold *through time*. The existence of states such as Cuba, Bolivia, North Korea, and Venezuela and the possibility for future democratic socialist and authoritarian experiments of this kind in the current global order affords space for comparative institutional claims about the advantages of predominantly liberal and social democratic market systems to be *continuously* challenged and to become more or less robust. Just as freedom of speech for those with minority opinions in a democracy sustains the conditions within which the majority may come to confirm or to revise its current opinions, so the existence of an institutional space for contrarian agents to implement alternatives to market capitalism and polycentric rule formation sustains the background conditions within which the case for the latter regimes may become more or less apparent.

The alternative to such meta-level polycentricity is meta-level monocentricity wherein data is produced by a process where only one institutional model or policy is implemented over a given time-period. The global enforcement of market capitalism or socialism or of a singular global approach to macro-economic management would be examples of such mono-centricity. Here, data production and evaluation would be limited to a process where outcomes before the institution or policy in question is introduced are compared with those after the institution or policy is introduced. Such methods are epistemically weak in comparison to metal-level polycentricity because they block the production of realised or observable counterfactuals. From questions of macro-economic governance, the size of the welfare state, through to the plethora of smaller scale legal and regulatory measures that seek to improve the performance of markets, states, and communal governance arrangements, it is often hard to discern which structures and measures may 'work' most effectively. What is clear, however, is that the capacity to make *any* of the empirically based evaluations that comparative institutions analysis requires, is facilitated wherever meta-level polycentricity prevails. In the case of macro-economic stabilisation for example, given the absence of clear counterfactuals it is notoriously difficult to discern which monetary, or fiscal policy measures 'speed' recovery from recessions compared with an absence of such measures, or if they do whether they generate negative unintended consequences elsewhere. Nonetheless, within a meta-level polycentric setting policymakers and citizens can at least observe the outcomes that follow the implementation of policies in different nations and jurisdictions drawing on rival macro-economic theories or none and try to tease out the implications for subsequent decisions. By contrast, if a *global* fiscal or monetary stimulus was introduced in response to a global depression and employment levels recovered after a number of years there would be no observed counterfactual to infer what would have happened to employment absent the stimulus. This problem may be mitigated if the 'trial period' for any institution or policy is sufficiently short such that counterfactuals might more easily be generated as would occur if a fiscal stimulus could be followed by equivalent periods without such stimulus. The longer the period over which a given institution or policy stays in place to the exclusion of others, however, the less robust will be any comparative empirical claims about the hypothesised effects of an alternative.

Climate Change Governance and Comparative Institutions Analysis

In view of the theoretical framework for comparative institutions reasoning, data production, and evaluation set out above how should one understand the character of the problem represented by anthropogenic climate change and the type of comparisons required to evaluate the possible institutional responses to it?

Most analyses of climate change understand the problem as a combined externality and collective action failure reflecting misaligned incentives. The production of greenhouse gases such as carbon dioxide has potential costs for both present and future generations that are not sufficiently factored into the decision calculus of those producing emissions because the latter do not pay a direct price to those who may be negatively affected by climate changes. Moreover, since the total level of emissions results from the actions of multiple producers and consumers, no individual actor, whether a person, corporation, or even a nation state, can have a significant enough effect on the overall level. The 'open access', nature of the global atmosphere is such that decisions to produce more, or less carbon dioxide will have little effect on the total produced so the optimal strategy for any self-interested party is to 'free-ride'. If other parties cease to emit, then any one decision to produce carbon will have minimal effect on emissions or on climate change. If, however, other parties continue emitting then any single decision to stop emitting will have a similarly minimal impact on total emissions.

Institutional Options for Climate Change Governance

Within this context, Coase lists four institutional options for responding to externality and collective action problems in general that are also the subject of comparative institutional claims in the climate change case.[11] The first involves relying on a market to identify and internalise costs by specifying property rights. This could arise through a 'bottom up' common-law process where a series of lawsuits against potential polluters generates a set of legal expectations about acceptable levels of emissions and property rights commensurate with these expectations. Alternatively, it could arise through a 'top-down' procedure wherein a judicial or administrative agency imposes a set of private rights to use or access the atmosphere. Either way, in this scenario the role of governance hierarchies would be a relatively limited one, confined to defining and enforcing the rights of a range of different atmospheric resource 'owners'.[12] Competing entrepreneurial bids for the assets of these owners based on varied estimates of their present and future value could then generate price signals, with the profits and losses of the respective entrepreneurs reflecting their relative success in forecasting present and future resource scarcities and market conditions. While such signals would contain many forecasting errors, they would limit the scope for such errors to be imposed systemically, relative to a situation where the value of resources is estimated centrally by state planners. In the context of climate change governance, however, these 'solutions' are usually considered unworkable with the indivisibility of the global atmosphere precluding the development of an effective competitive market process at the global level to determine the price and quantity of greenhouse emissions. This analysis would appear to underlie the suggestion in the Stern Review that climate change reflects the 'greatest market failure the world has ever known'.[13] The latter description might be contested on the grounds that the problem represents not so much the failure of a functioning market but the difficulty of *establishing* a market with the qualities often attributed to competitive processes. Whether one sees this problem as a 'market failure' or the 'inability to have a market', however, there are clearly pronounced obstacles to the operation of competitive market processes in determining the level of greenhouse emissions.

The second institutional option identified by Coase involves bringing decisions within 'decentralised hierarchies' such as firms, clubs, or local public bodies. Seen from a global level perspective the decentralised hierarchies in question would also include nation states and other sub-global associations or hierarchies such as the European Union. In the climate change context this arrangement roughly corresponds with the current global climate governance regime. Here, individual nation states or groupings introduce their own climate change policies through domestic governance hierarchies to discourage emissions and to induce technological change in the direction of lower emissions. These policies include carbon taxes and/or emissions trading systems. While these options are often described as 'market-based' policy tools because they rely in different ways on the use of price signals—they should not be confused with a 'free market' regime. Whereas in the latter both prices *and* quantities would be determined through competitive bargaining under carbon taxes prices are determined centrally by governments with quantities adjusting to these prices. With carbon trading by contrast, quantities are centrally determined but with prices then emerging through trade in the relevant quotas. As such, both models of climate 'pricing' are better understood as forms of 'market socialism' or 'administrative pricing' where the relevant prices/quantities are typically informed by econometric simulations rather than actual market processes. The case for these measures is that while addressing the 'market failure' resulting from the absence of a bottom-up pricing system for carbon they allow considerable scope for entrepreneurs and consumers to draw upon knowledge that may not be accessible to planners and regulators in deciding how to respond to the relevant signal. The major alternative to such 'price-based regulation' is the use of direct mandates that regulate or ban certain carbon-intensive production methods or that prescribe specific low-emissions production techniques. Sceptics of such mandates suggest that regulators lack the context-specific knowledge to discern which technologies are appropriate—knowledge more likely to reside with those in the industry concerned. For their supporters, however, there may be no substitute for regulatory prescription in contexts where there may not be time to wait for producers to adapt their behaviour in response to the more indirect pressures arising from price signals.

Under the current set of global arrangements, it is important to recognise that while nation states and groupings thereof are at liberty to determine what they see as the appropriate mix of climate policies—this includes the option of *not having any policy focussed on reducing emissions*. There is currently no global enforcement mechanism to mandate emissions reductions or climate policies of any sort with nation states remaining at liberty not to have any climate change policies at all. This is what the Bloomington school would understand to be a polycentric governance regime. An important element of the case for such a regime is that it allows for a process of reflexive learning across states as countries can discover the costs of alternative climate governance policies (such as price-based tools versus regulatory mandates) or none, to their respective economies by observing the effects of what other countries and associations do.[14] Correspondingly, the dispersal of decision-making authority reduces the prospects of systemic failure because there is no single centre of authority that can impose a policy error on a system wide basis. An important element of the case against such a regime, however, is that it fails to address the global level free rider or collective action problem and allows 'too much' scope for actors to refrain from making any reductions in their emissions and thus continuing to act in ways adverse to climate stability. The latter objection is analogous to arguments about the unworkability of market processes wherein the indivisibility of the global atmosphere precludes the internalisation of externalities at any level *below the global scale*. Those subscribing to the latter view maintain that climate change requires more centralised and potentially even authoritarian governance structures to properly address the problem and to discipline recalcitrant actors.[15]

The preference for more centralised and potentially authoritarian structures represents the third institutional possibility identified by Coase—that of direct central government regulation with the state acting as a single 'super-firm' or hierarchy that identifies relevant costs and imposes regulations via administrative fiat. In the context of climate change this approach implies either a global government with the power to enforce a set of rules pertaining to emissions on *all* nation states or a cartel of nation states acting in the manner of such a government. These rules might involve the imposition of a *single* global carbon tax or a single set of tradable global emissions quotas. Such options are consistent with maintaining a relatively high degree of polycentricity at the sub-global level as nation states would retain the freedom to set their own internal policies and to reduce emissions in such a way as to minimise their exposure to the centrally determined tax or quota price. Just as market-based tools adopted at the national level combine an element of centralised price determination with significant freedom for sub-national actors to experiment and to innovate in response to such prices, so the global enforcement of carbon taxes or tradable emissions quotas would leave considerable latitude for sub-global actors to experiment in calibrating their response to a globally determined carbon price. As such, they would exhibit many of the epistemic advantages claimed for 'price-based' forms of regulation. Equally, however, those who maintain that there is insufficient time to allow for market experiments and adjustments to take their course would favour the global enforcement of 'command and control' policy tools such as bans on fossil fuel extraction and mandated use of 'green' technologies.[16]

The major political economy objections to the adoption of either global carbon taxes/emissions trading or to a global command and control climate governance regime reflect the absence of anything resembling a global government with the necessary policymaking and enforcement powers, the huge transaction costs involved in creating such a regime, and the risks from creating a centralised power structure capable of inflicting systemic global level failures. The latter may arise from mistakes in forecasting the economic costs of climate change and of reducing emissions, and in the appropriate calibration of policy options. Alternatively, such failure could arise through institutionalised rent seeking and the abuse of monopoly governance power at the global scale.[17] While climate change might be understood as the greatest collective action problem the world has ever known, a collective action problem of not dissimilar proportions might face any global citizenry in attempting to ensure that the costs arising from forecasting error and/or rent seeking do not exceed the costs that may be associated with climate change itself.

The fourth and final option identified by Coase for addressing externalities is that of 'doing nothing'. This option might be favoured when the costs of the governance structures designed to address an externality exceed that of the externality in question—as might be the case if enforcement costs prove high relative to the size of the problem or where the effect of rent seeking or cognitive error by those operating the regulatory system is greater than that of the original externality. This position is rarely represented in the contemporary climate change debate. The closest approximation to it is, however, reflected in Lomborg's emphasis on adaptation, resilience, and technological advance through current models of economic growth.[18] On this view, 'excessive' attempts to limit emissions today may slow down the rate of economic growth and with it the resilience to future climate shocks that a higher growth rate today may afford. Moreover, attempts to 'force' the adoption of greener technologies 'too soon' may impede the process of capital formation that might generate cleaner and lower cost but as yet, 'unknown' technologies in the future. Critics of this perspective, however, maintain that it exhibits undue complacency in view of what they see as the scale of the threat represented by the effects of climate change. Others argue that it provides cover for vested interests and notably those of polluters in the fossil fuel industry to downplay the effects of climate change and to exaggerate the costs of more pro-active climate governance policies.[19]

The Problem of Comparative Evaluation

It should be evident from the above sketch that the comparative political economy arguments for or against these different governance regimes make implicit or explicit empirical hypotheses about the likely effects of the available institutional alternatives. Recall that from the standpoint of comparative political economy institutions should not be judged against an external or exogenous ideal standard. Notions of what constitute relatively better arrangements must combine theoretical speculation with empirical *observation of the effects of the relevant alternatives.* If it is not known *ex ante* which institutional arrangements are best placed to identify and to internalise the costs of greenhouse gases, then it follows that some sort of trial and error method, where different theories or conjectures about the effects of governance alternatives are tested, will be needed to discern the direction in which political economic institutions should be modified. To establish that climate change is 'the greatest market failure the world has ever known', or that polycentric institutions allow 'too much' scope for free-riding, requires empirical evidence on how the outcomes people experience under markets or polycentric institutions compare to those generated under an alternative. Similarly, to establish that the risks from a global climate governance monopoly outweigh the risks from insufficient regulatory action under a polycentric regulatory order or those arising from a 'do nothing' scenario requires a process through which the effects of these possibilities can be observed and compared. To adapt a phrase from James Buchanan, there are no external 'data' against which to compare the costs and benefits of alternative climate governance institutions *outside of a process that allows for the emergence of their comparative effects*—i.e., that allows for the production of the relevant data.[20]

Yet the question remains as to how the relevant empirical observations are to be generated and evaluated in the climate change context. What was previously referred to as meta-level polycentricity would involve allowing the different approaches to climate change—in effect all four of the governance options outlined by Coase—to be tried in parallel with data on the comparative effects of the different arrangements emerging from an ongoing process of institutional contestation and competition. When a governance problem occurs at the sub-global level, such as how best to value and to manage forests or regional water resources such meta-level polycentricity allows for the production of comparative data about the effects of managing the relevant resources by leaving them in an open-access state (doing nothing), relying upon private ownership, public ownership, communal or mixed governance regimes, and between more, or less polycentric and monocentric structures. It should, however, be readily apparent that such meta-level polycentricity is simply not viable as a form of comparative data production and evaluation in the case of climate change. The indivisibility of the global atmosphere means that the effects of the different institutional responses cannot adequately be separated from each other such that any 'contamination effects' might be sufficiently removed. If a group of nations pursues a radical emissions reduction program through carbon taxes, emissions trading or command and control regulation while another group pursues a 'do nothing' growth and adaptation agenda there will be no way to discern how far reductions in emissions have modified the effects of climate change and whether the costs associated with these reductions are outweighed by the benefits in a setting where other nations are at liberty to engage in activities that may continue to disturb the climate.

If meta-level polycentricity cannot generate the necessary comparative institutional data neither however can such data be generated through meta-level monocentric evaluation. Before and after evaluations can and have provided useful data on *some* of the comparative costs of various environmental policy tools. Comparative studies have, for example, shown that tradable permit schemes to reduce Sulphur Dioxide emissions significantly lowered costs relative to the 'command and control' policy tools that preceded them.[21] What such methods cannot do, however, at least not in the climate change case, is to compare the effects on human welfare of different attempts to forecast the relative value of reducing emissions in comparison to some alternative strategy—such as one based on adaptation rather than mitigation—or of different combinations of adaptation and mitigation. Given that many of the possible effects of climate change may occur in the relatively distant future the period between which the implementation of any governance regime and the forecasted effects this may have both on socio-economic conditions and on climate change may be too great to allow for comparison with the effects of alternative institutions, policies, and forecasts over an equivalent period. If a regime of global carbon taxes or emissions trading is implemented for a period of say fifty years, then it will not be possible to assess effectively what would have happened under an alternative. In other words, any global level climate governance regime—*including the decision not to have such a regime*—will block the production of *actualised* as opposed to hypothetical data about the opportunity costs associated with an alternative regime. Equally, however, actualised data about the relevant institutional opportunity costs cannot be generated if climate governance regimes are repeatedly overturned. There would, for example, be little point in pursuing a regime of stiff carbon taxes for several years only then to abolish them. While such a process might reveal the short-term economic costs and benefits of the policies concerned it could not generate a comparison of these costs with the future costs that may or not prove to be associated with the effects of climate change.

Modelling to the Rescue?

One response to the evaluative problem set out above is to invoke the use of econometric modelling and other simulation techniques to comparatively assess the effects of climate change and alternative institutions and policies. Such techniques underpin the extensive modelling and data collection exercises that have been conducted by multiple academics and policy analysts across the world in attempts to estimate the 'social cost of carbon'.[22] As things stand, there are significant variations in these modelling estimates with some pointing towards relatively minor negative effects from climate change and others much more significant effects. Notwithstanding these differences it could be argued that the 'consensus position' within this literature which implies a case for moderate carbon taxes or emissions quotas, represents the 'best available estimate' of the possible costs associated with alternative governance scenarios.[23] At the very least, this would imply rejecting 'do nothing' or 'business as usual' policies and opting for some sort of carbon pricing regime, with or without a global enforcement process. From the standpoint of comparative political economy analysis, however, the problems with this line of reasoning are threefold.

First, existing efforts to model the costs of climate change and of climate change policy tend to treat the possible effects in isolation from multiple other interacting policy domains, variables, and questions of political economy. Political economic systems are, however, best understood as complex systems or 'entangled political economies' where interventions in one part of a system—such as a carbon tax—can have multiple unintended consequences elsewhere in that system.[24] Within this context, it is rarely if ever considered in econometric simulations, whether policy attempts to price carbon emissions will prompt responses from those who will be subject to their effects in ways that may counteract any benefits the respective policies may have. Consider in this context the current efforts by the UK government to achieve 'net zero' carbon emissions. The significant rise in the price of energy arising from this policy commitment, which accounts for a quarter of household energy costs, is now prompting multiple demands for the introduction of price controls in the retail energy market and populist demands for the nationalisation of major energy suppliers—policies that if they were pursued might have significant negative impacts on the efficiency of the energy sector and on the wider economy.

Second, even if attempts were made to factor these political economy variables into the relevant modelling scenarios, it is far from clear that such techniques can do so reliably. Econometric techniques may be a useful compliment to the decisions of individuals and organisations in their day-to-day planning and attempts to forecast *near term* possibilities because there are *some* regularities arising from the fact that the decisions people make in the immediate future typically grow incrementally out of their previous decisions. Given radical uncertainty about the factors that will shape the contours of the more distant future, however, longer term econometric modelling maybe of little use because reliable forecasts require detailed knowledge of all the factors—such as the contribution to welfare of technology and resources not yet invented or discovered, and how these will be affected by alternative climate changes policies in the future, as well as the political economy considerations discussed above—that will shape the future *in advance of their emergence.*[25] At the very least, therefore, the epistemic credentials of modelling exercises that purport to model the costs and benefits of alternative institutional and policy scenarios for decades if not centuries in advance are highly questionable.

Third, and most important, the relative value of econometric modelling as a guide to decision-making cannot itself be ascertained outside a context that allows either for meta-level polycentricity or meta-level mono-centricity to generate the necessary 'data' about its comparative effectiveness. While the case for econometric techniques does not depend on their optimality when judged against some external or ideal standard, establishing that they are *relatively better* guides to decisions than alternatives in a particular setting (such as actual market prices), requires some kind of 'real world' comparative process. As noted earlier, attempts to price greenhouse emissions through econometric simulations are a form of 'administered pricing' or 'market-socialism' and while huge amounts of people and resources might be devoted to informing these exercises it must be remembered that similar volumes of data were generated by the former socialist economies in evaluating and reconfiguring their economic plans and forecasts. The epistemic failure of these plans and forecasts was not, however, revealed by modellers themselves, but by a meta-level setting of polycentricity that exposed the relatively poor performance of economies reliant on economists and planners to simulate prices and quantities in comparison to those that relied on actual though imperfect markets. Such meta-level comparisons cannot however be operative in the climate change case. Neither can the comparative merits of reliance on econometric techniques be established through meta-level mono-centricity because the timescales necessary to evaluate their relative efficacy in forecasting the effects of climate change and of policies to counteract it, are too great to allow for comparison with those that might be generated by some alternative procedure or non-procedure.

History to the Rescue?

If econometric simulations do not adequately address the problem of comparative institutional evaluation an alternative response to the problem outlined here might be to suggest that in deciding the appropriate response to climate change citizens and policymakers alike may draw on the lessons of history. While historical comparisons are far from ideal, they have the advantage of drawing on 'real-world' experiences that may help to at least narrow the choice between different institutional and policy options. This type of reasoning might for example highlight the dangers of a 'do-nothing' approach to climate change by drawing on the historical experience of previous climate change events to illustrate the potential scale of the threat/s they have represented. Similar lessons might also be invoked in this regard to rule out the more draconian calls for centralised global planning and the widespread adoption of command, and control measures. If history shows us that measures of this kind when implemented at the sub-global level have impoverished societies and have led to unaccountable and indeed tyrannical systems of elite rule, then surely any approach that proposes similar controls at the global level to address climate change should be excluded from serious consideration.

History may well furnish important lessons for comparative institutional analysis, but it is far from evident what these lessons might be in the climate change context. One reason for this is that climate changes referred to in the past were not significantly if at all human induced and they took place in the context of significantly poorer societies that had none of the resilience that has arisen through centuries of economic growth. The latter has been responsible for anthropogenic climate change, but the fruits of such growth may also provide the means to better cope with climatic shifts than has historically been the norm—irrespective of whether these are 'natural' or have been induced by greenhouse emissions. It is not, therefore, evident how history can help to discern the relative merits of a view that favours relatively unhampered economic growth in the near term as the best insurance policy against the possible effects of future climate changes, and those who advocate potentially costly climate mitigation measures now as providing the best such insurance. Neither is it apparent how history can determine whether global central planning or direct regulation of greenhouse gases should be discarded as an institutional option. Those favouring such measures could argue that what they propose is not equivalent to Soviet style planning of *entire* economies but to central planning *on one dimension* alone—albeit a very significant one. Given what many perceive as the unique scale of the climate threat such one-dimensional planning might be justifiable—and especially so when recognising that at the sub-global level nation states and other jurisdictions routinely engage in command and control methods including bans on the use of substances such as mercury that they deem to be excessively dangerous.

Like Cases to the Rescue?

A final response to the evaluative dilemma set out here might be to invoke examples of like cases from which to draw comparative institutional lessons. This is not the first time that the world has been confronted with problems that share some of the features of the climate governance challenge and so perhaps institutional lessons might be drawn from these experiences. The threat to the ozone layer arising from the widespread use of chlorofluorocarbons in refrigeration and aerosols might, for example, be understood as a similarly large-scale global collective action dilemma. In this case, coordinated global policy action organised through the Montreal Protocol led to a significant reduction in CFC emissions that has allowed for at least a partial reconstitution of the ozone layer, while the costs of reducing the use of CFCs have transpired to be *less* than originally anticipated. In this instance, few would wish to argue that a comparative assessment of institutional alternatives has not been possible. Given the seriousness of the threat to the ozone layer the regulatory actions that were taken at the global and sub-global level to address this issue must surely be judged as better options than either a 'do nothing' approach to the externality, or an attempt to rely on price signals to internalise the relevant costs.

From the standpoint of this chapter, however, a comparative institutional endorsement of the response to ozone depletion cannot so easily be transferred to the problem of evaluating the possible responses to anthropogenic climate change. In the case of CFC emissions relatively cheap alternatives for refrigeration and aerosols were already available such that there was never a prospect that shifting to these would generate significant short-medium term economic costs that might outweigh the need to prevent the clear and severe negative consequences that would result from the destruction of the ozone layer. The risks of skin and other cancers arising from ozone depletion in the relatively near term, for example, constituted a transparent motivation for political action. The danger presented by a failure to reduce CFC use was evident and the costs of moving to alternatives were relatively low and well known. Though some poorer nations required temporary assistance to sustain these costs, in a context of generally rising incomes they did not impact on a sufficiently large share of economic activity to make these a long-lasting or expensive commitment. As such, in the ozone depletion case, the conditions were relatively close to those where monocentric 'before and after' evaluations may enable adequate empirical comparisons of alternative institutions and decisions to be made.

In comparison to CFCs, fossil fuel use is, however, currently much more integral if not fundamental to the economic model of richer and poorer countries alike and there are no cheap and available substitutes– or at least none that do not themselves come without potentially significant economic and environmental costs (such as nuclear power). Where CFC use was concentrated heavily in a few industries that could relatively cheaply be compensated for transition costs, the use of fossil fuels in production and consumption permeates the entire economic structure and correspondingly many of the proposals for reducing emissions involve significant economic and regulatory costs across multiple different sectors. As Adler[26] puts it, 'buying new refrigerators and retrofitting car air-conditioners may be expensive and inconvenient but it hardly compares to the sorts of measures contemplated to reduce greenhouse emissions'.

Whether fossil fuel substitutes might become available at a price less than the costs of future climate changes is unknown in part because the nature of the threat is also much less evident. Most climate change scenarios *do not* see the problem as an existential threat but one where, depending on the level of warming, there may be relatively marginal reductions in economic growth compared to scenarios without climate change.[27] Owing to the significant number of excluded and unknown variables associated with the relevant modelling efforts, forecasts about the potential socio-economic consequences of climate change decades, if not centuries from now are thus far more tenuous than those that were associated with the imminent threat of ozone depletion. Moreover, whereas the effects on the ozone layer of reductions in CFC emissions could be tracked over a relatively short period (a decade or so), the timeframe within which it might become possible to detect whether reductions in greenhouse emissions have had any effect on the trajectory of climate change at all will be much more difficult to assess. In these circumstances, there is no obvious way to discern which institutions and policies are best placed to make the trade-off between the value of current economic growth against predictions about the possible effects of climate changes in the relatively distant future.

Of course, one objection to all the arguments advanced thus far might be to say that uncertainty is prevalent in *all* human decisions and that the need to negotiate such uncertainty is no different in principle in the case of climate change governance. This would, however, be to miss the central point that this chapter has sought to make. The point is precisely that in many decision-making contexts the way to manage radical uncertainty is to hedge one's bets, to fracture or to divide decision-making authority in a manner that allows for comparative trial and error data production, that facilitates reflexive learning across decision-making units, and that reduces systemic risk—and this is what meta-level polycentricity allows for. Such an option is simply not available in the climate change context and consequently no institutional option can be said to reduce the risk of systemic error or failure. In this sense, the problem of evaluating institutional responses to climate change might be considered of an even greater complexity than those involved in questions of national security or of decisions to go to war. While decisions, or non-decisions, must be made to judge for example whether nuclear armament increases or reduces the risk of conflict, or whether the 'war on terror' increases, decreases or makes no difference to the level of terrorism, in the absence of counterfactuals such decisions can made with little confidence.

Conclusion

The foregoing analysis has emphasised the importance of comparative institutions reasoning in political economy and of judging institutions against 'real world' alternatives. While theoretically informed judgements about the capacity of different institutions to address problems of incentive compatibility and the cognitive limitations of decision-makers are valuable, the argument here has suggested that the empirical magnitude of the hypothesised effects and their implications for institutional choice cannot be known outside a context that allows for institutional alternatives to be tried and tested against one another. In the case of anthropogenic climate change, however, the conditions required for such institutional contestation to produce the necessary data may not exist. For climate change at least, the claims of comparative political economy cannot therefore be considered as empirically robust, and indeed it is a primary implication of this chapter that any such claims should be considered as no more than theoretical speculation. As such, the analysis offered in this chapter is strictly diagnostic and *not prescriptive*. It suggests that citizens, academics, and policymakers alike must decide their response to the climate change dilemma in a fog of comparative institutional ignorance. For some, wanting or having to make decisions about climate change governance, this diagnosis may be greeted as an exciting opportunity to take bold experimental steps, and to partake in a voyage of exploration into the unknown. For others, however, it may well be considered to represent the ultimate tragedy of anthropogenic climate change.

Notes

1. Consistent adherents of the omniscient rational choice assumptions that underpin the Chicago perspective have recognised this problem but only in such a way as to render *any* kind of comparative institutions reasoning redundant. Wittman, for example, argues that *any* existing institutions whether based on markets or some form of government intervention or regulation should be considered efficient because if they were not, rational, well-informed actors would already have changed them. In other words, whatever is, *must be efficient*. See Donald Wittman, *The Myth of Democratic Failure* (Chicago, IL: University of Chicago Press, 1995).
2. Ronald Coase, "The Problem of Social Cost," *Journal of Law and Economics* 3, no. 1: 1–44; see also Ronald Coase, *The Firm the Market, and the Law* (Chicago, IL: University of Chicago Press, 1989).
3. Harold Demsetz, "Information and Efficiency: Another Viewpoint," *Journal of Law and Economics* 12, no. 1 (1969), 1–22.
4. J. Buchanan, and G. Tullock, *The Calculus of Consent* (Ann Arbor, MI: University of Michigan Press, 1962).
5. Elinor Ostrom, *Governing the Commons* (Cambridge: Cambridge University Press, 1990).
6. Friedrich Hayek, "The Meaning of Competition," in *Individualism and Economic Order*, ed. Friedrich Hayek (Chicago, IL: University of Chicago Press, 1948). Friedrich Hayek, "Competition as a Discovery Process," in *New Studies in Philosophy, Politics, Economics and the History of Ideas*, ed. Friedrich Hayek (London: Routledge, 1978).

7. The phrase 'robust political economy' has been used to describe this emphasis on motivational and cognitive symmetry—see, for example, Mark Pennington, *Robust Political Economy* (Cheltenham: Edward Elgar, 2011). See also Neil Komesar, *Imperfect Alternatives* (Chicago, IL: University of Chicago Press, 1994). In similar vein Demsetz, *Information and Efficiency*, refers to avoiding the 'nirvana fallacy'.
8. Gordon Tullock, *Rent Seeking* (Cheltenham: Edward Elgar, 1994).
9. Don Lavoie, *Rivalry and Central Planning* (Cambridge: Cambridge University Press, 1985); Don Lavoie, *National Economic Planning: What is Left?* (New York: Ballinger, 1985).
10. Ostrom, *Governing the Commons*; Ostrom, *Understanding Institutional Diversity*.
11. Coase, *The Problem of Social Cost*.
12. For example, Graham Dawson, "Austrian Economics and Climate Change," *Review of Austrian Economics* 26 (2013): 183–206.
13. N. Stern, *The Economics of Climate Change* (Cambridge: Cambridge University Press, 2007).
14. Elinor Ostrom, *A Polycentric Approach for Coping with Climate Change* (World Bank, 2009). See also, Daniel Cole, "Advantages of a Polycentric Approach to Climate Change Policy," *Nature Climate Change* 5 (2015): 114–118.
15. Ross Mittiga, "Political Legitimacy, Authoritarianism and Climate Change," *American Political Science Review* (2021): 1–14.
16. Mittiga, *Political Legitimacy, Authoritarianism and Climate Change*.
17. For a discussion of these possibilities see Ostrom, *A Polycentric Approach for Coping with Climate Change*.
18. Bjorn Lomborg, *False Alarm* (New York: Basic Books, 2020). Lomborg's position is more pro-active than the phrase 'do nothing' implies. He advocates technological advance and resilience to climate change through growth but also supports the subsidisation of cleaner technologies as part of that process. A purist 'do nothing' position would presumably maintain that technology should be allowed to evolve in the absence of policy-induced attempts to favour some technologies over others.
19. For example, Peter Frumhoff, Richard Heede, and Naomi Oreskes, "The Climate Responsibilities of Industrial Carbon Producers," *Climate Change* 132 (2015): 157–171. Note, while Frumhoff, Heede and Oreskes may be correct in emphasising the scientific consensus on the phenomenon of anthropogenic climate change—for a contrary view

see Warren Pearce, Reiner Grundmann, Mike Hulme, Sujatha Raman, Eleanor Hadley Kershaw and Judith Tsouvalis, "A Reply to Cooke and Oreskes on Climate Science Consensus Messaging," *Environmental Communication* 11, no. 6 (2017): 336–339—there is nothing resembling an equivalent consensus in estimating the socio-economic costs of alternative climate change scenarios and policies.

20. James Buchanan, "Order Defined in the Process of its Emergence," *Literature of Liberty* 5, no. 4 (1982). Buchanan is making the point here, that in market settings individuals do not maximise utilities that are captured by independently existing utility functions. Rather, market participants do not in fact know what their choices and utilities will be until they enter a process that presents different possibilities to them. This is a social constructionist argument which implies the absence of external or objective data with respect to costs and benefits—in short there are no costs and benefits outside of the social processes that produces or enables comparisons between different production and consumption possibilities—and by implication here, different institutional possibilities.
21. Denny Ellerman, Paul Joskow, Richard Schmalansee, Juan-Pablo Montero, and Elizabeth Bailey, *Markets for Clean Air: The US Acid Rain Program* (New York: Cambridge University Press, 2000). Note there is some dispute here about the extent to which emissions trading per se, as opposed to other changes in the policy environment in the United States were responsible for this success. On the one hand, the program allowed the substitution of low Sulphur western coal, for high Sulphur eastern coal—and at the same time transport de-regulation reduced the costs of shipping coal across the United States. It seems, however, that relatively little trading in sulphur dioxide permits took place. I owe this point to Jonathan Adler.
22. Mark Budolfson, "The Social Cost of Carbon, Humility, and Overlapping Consensus on Climate Policy," this volume.
23. William Nordhaus, *The Climate Casino* (New Haven: Yale University Press, 2013).
24. Richard Wagner, *Politics as a Peculiar Business* (Cheltenham: Edward Elgar, 2016).
25. On the problems of econometric forecasting which continues to be of relevance see George Shackle, "Keynes and Today's Establishment in Economic Theory: A View," *Journal of Economic Literature* 11, no. 2 (1973): 516–519.

26. Jonathan H. Adler, “Ozonegate and Climategate,” The Volkoh Conspiracy Blog (December 7, 2009).
27. Nordhaus, *The Climate Casino*; Peter Cox, Chris Huntingford, and Mark Williamson, “Emergent Constraint on Equilibrium Climate Sensitivity from Global Temperature Variability,” *Nature* 553 (2018): 319–322.

The Social Cost of Carbon, Humility, and Overlapping Consensus on Climate Policy

Mark Budolfson

Introduction and Overview

Social cost of carbon (SCC) estimates are the primary guide to carbon pricing policy proposals and are also commonly used along with other models to estimate the magnitude of greenhouse gas (GHG) emissions reductions that it would be best for nations and the world to make in the coming decades in order to mitigate climate change.

At first glance, it may seem that an SCC-based approach to climate policy (e.g., an economy-wide carbon tax calibrated to the SCC) presupposes a set of very controversial assumptions, especially about what detailed knowledge regulators have about the impacts of climate change and what the proper role of government and policy is in responding to those impacts. At worst, it may initially appear that an SCC-based approach assumes an omniscient and omnibenevolent social planner who

M. Budolfson (✉)
Rutgers University, New Brunswick, NJ, USA
e-mail: markbudolfson@gmail.com

J. H. Adler (ed.), *Climate Liberalism*, Palgrave Studies in Classical Liberalism,
https://doi.org/10.1007/978-3-031-21108-9_14

can be relied upon to perfectly implement a perfectly known regulatory "solution" to the problem of climate change (and solve all other inefficiency and equity problems in the background)—all of which would be an implausible set of assumptions if those were truly a presupposition of the SCC-based approach.

However, as I explain in what follows, the SCC-based approach need not have these problematic presuppositions, and in fact SCC models can provide the best guide to climate policy when implemented in a way that incorporates a healthy dose of humility. In the next section I begin by briefly explaining the SCC-based approach, and then in the section "A Modest Carbon Tax Derived from SCC Estimates as an Alternative to a Pigouvian Tax," I explain why it is compatible with humility about our knowledge of the impacts of climate change, humility about the incentives of regulators and other actors in society, and humility about the limits of government.

I then explain in the section "Overlapping Consensus, Ecumenical Climate Policy, and the SCC-Based Approach" why the SCC-based approach can be used in a way that is ecumenical between the wide range of reasonable but incompatible views about the proper goals of government and policy, ranging from mere market-efficiency aimed views, to utilitarian views, to rights-based and other deontological views, to libertarian views, to virtue ethics views, and others. Beyond this, I suggest that the SCC-based approach can help us find an overlapping consensus on a particular climate policy given the range of reasonable but incompatible normative views endorsed in diverse societies, where these views in many cases will have incompatible implications about what climate policy would be ideal. This may have important implications for what climate policy is politically justifiable given the fact of reasonable pluralism. It also may have implications for what is justifiable according to some specific views, especially views such as deontology and/or libertarianism that see the mere promotion of aggregate social welfare as an insufficient justification for the coercive policy, and require instead broad-based consent to such coercion as necessary for justification. In this way, I argue that the SCC-based approach is compatible with liberal (and classical liberal) conceptions of the proper role of government. In addition to these normative considerations, identifying an overlapping consensus may also have practical political importance, as it may help identify a strategic coordination point for climate policy around which a successfully coalition may be built.

Finally, in the section "Additional Worries About the SCC-Based Approach," I consider a number of more specific objections to the SCC-based approach that are especially prominent in contemporary discourse and policy debates. These include several arguments that the SCC-based approach systematically recommends too much mitigation, as well as several arguments that it systematically recommends too little mitigation. I also consider an argument that it recommends the wrong mechanisms for emissions reductions. I suggest that even if we agree that all of these objections to existing SCC-based analyses have an important kernel of truth, they do not undermine arguments that the correct response to climate change still involves substantial emissions reductions to be achieved via policy, and that the best methods for deciding the magnitude of those reductions still depend essentially on the SCC-based approach. Thus, even those who agree with these important worries do not thereby have a good reason for wholesale rejection of policy guided by SCC estimates, and in fact they should still find action-guiding value in those estimates as one of the best available tools for deciding what we should do about climate change.

The Social Cost of Carbon

The Social Cost of a Pollutant: The Basic Idea

Before explaining in detail the methods used to estimate the social cost of carbon (SCC), it may be useful to begin with a simple illustration of the basic idea of the social cost of a pollutant. To illustrate the basic idea, imagine a hypothetical world like ours except that instead of fossil fuels industrial activity often relies on intoxicating fuels. Specifically, suppose that these intoxicating fuels function exactly like fossil fuels in industrial activity, but fortunately do not have greenhouse gasses as a side effect of their combustion. Unfortunately, they instead emit an intoxicating gas that quickly spreads evenly through society. Suppose that this intoxicating gas has little effect in small concentrations but is a big problem in aggregate concentrations that result from the large sum of all emissions of the intoxicating gas across society. Suppose further that after careful observation over many decades, we can see a clear increase in accidents and a clear reduction in worker productivity, and social and natural scientists have identified this intoxicating gas as the main culprit beyond a reasonable doubt.

To make this imaginary case simple and clear (but also disanalogous to the case of climate change, which involves greater uncertainty), suppose that scientists can predict fairly precisely the magnitude of intoxication that will result from higher and lower concentrations of the pollutant. Suppose further that economists and other researchers can also make fairly precise, high-confidence estimates of the additional deaths due to accidents that should be expected as a function of higher and lower concentrations of the gas, as well as the greater and lesser loss of GDP we should expect as a function of higher and lower concentrations of the gas (given the loss of productivity due to well-understood dynamics of intoxication).

Given all of these suppositions, it is easy to see how in this hypothetical case we could make a good quantitative estimate of the harm to people now and into the future of additional emissions of the intoxicating gas. In particular, the previous paragraph supposes that scientists can reliably and precisely identify "damage functions" along at least two dimensions of harm: first, the quantity of additional deaths from accidents that would result from an additional unit of emissions, as well as, second, the quantity of lost GDP that would result from an additional unit of emissions. Following the relevant terminological conventions—and in the sense of "social cost" relevant to this paper—we can call the sum of these harms from additional emissions of the gas the social cost (i.e., the cost to society) of additional emissions of that gas.

Note how useful it can be to this society to have an estimate of the social cost of intoxicating gas emissions. Among other things, such an estimate would tell them quantitatively how bad additional emissions are in terms of their net harm to society, as well as conversely how much they would benefit from reducing emissions. This would then be useful in deciding what specific responses to the problem might be better and worse. For example, if people in this hypothetical scenario were considering a policy to reduce emissions by 50% over a decade that was known via the social cost estimate to imply avoided GDP loss of $N billion a year as well as M lives saved, then such a policy would be better than doing nothing insofar as they could be sufficiently confident that it could be implemented successfully and cost much less than $N billion, and not be a bad idea for other reasons. Of course, correctly estimating the cost and side effects of a specific policy is a much more difficult challenge that goes beyond calculating the social cost, and this will be further discussed below. The current point is merely to illustrate that the social cost of a pollutant is an easy-to-understand fact that can be investigated using familiar methods from science and economics, that in some cases it could be useful and feasible to estimate quantitatively, and that it is a proper subset of the overall policy equation (which must also take into account cost and feasibility of policy, broadly construed, along all other dimensions, as well as other factors). With this simple illustration of the basic idea of the social cost of a pollutant in hand, I turn next to explaining the methods used to estimate the social cost of carbon.

The Social Cost of Carbon (SCC), and Estimates of the SCC

The social cost of carbon (SCC) is the net harm to everyone now and into the future of an additional unit of carbon emissions. As such, an estimate of the SCC aims to capture the cost to society of additional CO_2 emissions and, conversely, the benefit to society of reducing those emissions. By convention, the SCC is estimated for one additional ton of CO_2 emitted in a particular year, and is monetized in present-day dollars.

Methods of estimating the SCC are analogous to the hypothetical example above, where the main task is to account for the economic, health, and other non-market impacts of the pollutant now and in the future. Estimating the SCC is much more complex than the example above because there are many more dimensions along which health and living standards are impacted by carbon emissions and these impacts are much more uncertain, especially in long-term timescales. In outline, existing methods begin by estimating the climate impacts of the emission of one additional ton of CO_2 (i.e. the impacts on temperature, sea level rise, etc.) and then estimate the economic, health, and other non-market impacts of those climate impacts for each year into the future (e.g. both positive and negative impacts, such as increased air conditioning expense, lower heating expense, loss to long-run GDP, loss from increased flooding, increased mortality and morbidity from heat stress, malaria, etc., decreased mortality and morbidity from frigid winter temperatures, etc.). They then monetize all of those disparate economic and non-market impacts for each year into the future so that a single net monetized damage can be calculated for each future year before calculating the SCC as the present value of the sum of all of those future net yearly damages (where the present value calculation discounts future damages in a principled way, e.g. taking into account interest rates, the anticipated wealth of future people, and [optionally] pure time preference).[1]

In this way, an SCC estimate aims to capture in clear quantitative terms the cost to society of additional CO_2 emissions and conversely the benefit to society of reducing those emissions. Note that each step in this analysis is more multi-dimensional and more uncertain than the simple example in the previous section. Nonetheless, many of the most important components of the SCC analysis are similarly straightforward (e.g. estimating heat-related mortality as a function of higher temperatures, estimating the impact on long-run GDP of heat-related productivity loss, estimating the net cost of heating and cooling expenses as a function of higher temperatures) and are tractable questions that can similarly be productively investigated by scientists and economists, albeit with wider bands of uncertainty.

The SCC-Based Approach to Climate Policy

Given SCC estimates, it is a further substantive question what specific climate policies should or should not be recommended given those estimates. An SCC-based approach to climate policy uses SCC estimates in an important way as a basis for climate policy recommendations. One way of doing this is to simply derive climate policy from SCC estimates using a particular "formula" for policy.

For example, the most common and simple way of deriving climate policy from SCC estimates is also the leading example of market-based environmental policy in many economics and policy textbooks: namely a Pigouvian carbon tax, in which a carbon tax is imposed each year into the future equal to the SCC estimate along the optimal emissions reduction trajectory (where the marginal abatement cost equals the marginal social cost of emissions)—and where this serves as the only climate policy.[2] Practically speaking, the goal of a Pigouvian tax is to force polluters to pay the true cost to society of their emissions while creating incentives across society that lead to a new outcome in which emissions of that pollutant are reduced to the optimal level. This is the optimal level in the sense that if emissions were reduced any further via a higher price there would be incremental economic costs to society that would outweigh the incremental benefits of the avoided pollution, and if those emissions were allowed to be any higher via a lower price, there would be incremental costs to society in terms of additional pollution that would outweigh the incremental economic benefits of allowing the additional pollution. Investment in the development of new clean technologies is thereby also incentivized in a way that reflects the true costs and benefits to society of those investments—and additional policies are inadvisable because they would only create unnecessary further distortions. In economic theory, this is a first-best regulatory response to the negative externality of carbon pollution, which has the virtue of reducing total emissions to an optimal level in the way that has the lowest possible cost to the economy (given standard textbook assumptions).[3]

Note that while the Pigouvian carbon tax is an instance of the SCC-based approach, it is only one among many different possible ways of deriving climate policy from an SCC estimate. Specifically, the Pigouvian tax approach adds the following controversial assumptions that go beyond the basic commitments of the SCC-based approach: namely that (i) the one and only instrument of climate policy should be emissions pricing, (ii) a tax is the best method for pricing (as opposed to e.g. cap-and-trade), (iii) the tax should be set at a level that would maximize economic benefits according to central estimates of the SCC, and (iv) side effects, background inefficiencies, and inequities from other sources can be ignored or should be dealt with by other policies.

Each of these assumptions is controversial and point the way toward possible alternatives to a Pigouvian tax. In the next section, I'll highlight one particular alternative that should be more widely discussed, especially because it illustrates the ability of the SCC-based approach to mitigate important worries about a Pigouvian tax, especially worries about assumptions (iii) and (iv) above.

A Modest Carbon Tax Derived from SCC Estimates as an Alternative to a Pigouvian Tax

If one believes that climate change is a deep problem for society and understands the concept of the social cost of carbon (SCC), then one is committed to thinking that the SCC is not zero—in other words, that there is indeed some net harm done by additional carbon emissions that could at least in principle be quantified. If one also believes that science and economics can help us estimate the SCC but that large uncertainties and deep methodological challenges are involved in such estimates, then one basically agrees with mainstream climate economics, which reports the social cost of carbon with the disclosure of many methodological challenges and increasingly focuses on stochastic methods to more precisely estimate the contours of our wide band of uncertainty about the true value of the SCC given the evidence.[4]

With this uncertainty in mind expressed as confidence intervals for the true value of the SCC, we can consider a simple SCC-based alternative to a Pigouvian tax: Instead of a carbon tax set equal to the central estimate of the SCC (which the Pigouvian tax does as a consequence of its assumption (iii) above that benefits are to be maximized in expectation), a carbon tax could instead be set to a lower level (a fraction of the central estimate), corresponding to a higher degree of confidence that the true value of the SCC was at that lower level or higher. This might be particularly attractive to those who agree that climate change is harmful on net, and thus the SCC is positive, and agree that collective action is necessary to mitigate this harm, but who are also concerned about the limits of our knowledge and the limits of government to estimate the best policy response and have the right incentives to do what is best.

To give a name to this alternative approach and to give it concreteness for our subsequent discussion, let's call it the Modest Carbon Tax, which (merely for ease of exposition) we can stipulate more precisely to be a carbon tax set equal to the point at which SCC estimates are (let us stipulate) 95% confident that the true value of the SCC is at that level or higher.[5] The arbitrary choice of 95% illustrates how this alternative approach can deliver a precise concrete formula for deriving policy recommendations from SCC estimates. Of course, this precisely stipulated value is just one member of a family of recommendations that could result from the choice of different confidence levels. To see the more general family, note that any choice of confidence threshold above 50% yields an SCC-based alternative to the Pigouvian approach.

The more general conceptual point that uncertainty analyses of social cost can yield more modest policy recommendations than the central estimate simply by choosing a higher confidence threshold is illustrated by Fig. 1.

One thing to note about the Modest Carbon Tax (and other members of the more general modest policy family that it represents) is that it

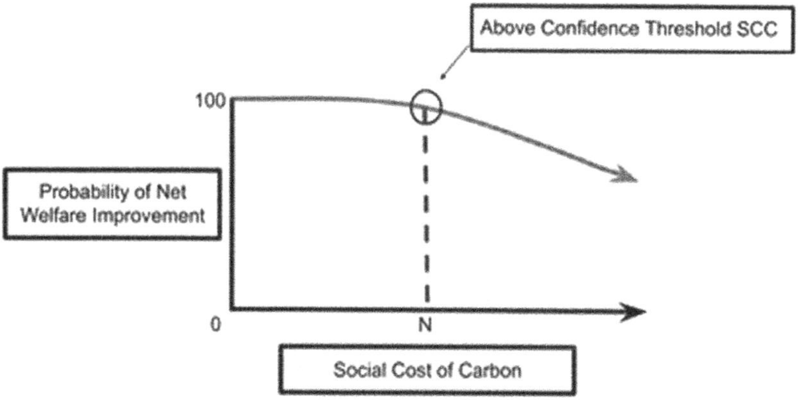

Fig. 1 Probability of net welfare improvement as a function of policy implementing specific social cost of carbon estimates

is indeed an SCC-based approach, based on the very same SCC estimates as the Pigouvian approach and equal to the Pigouvian approach in simplicity, but yet it provides a clear and substantive alternative to the Pigouvian approach. It rejects assumption (iii) of the Pigouvian approach that policy should aim to maximize welfare, and it presumably faces less worry in connection with (iv), as a lower carbon tax would be a less powerful intervention along the same vector, and thus would have less troublesome implications than a higher carbon tax regarding negative side effects, interactions with background distortions and inequities, etc.

There are a number of important upshots of this simple point. Perhaps the most important for our purposes is that the Modest Carbon Tax shows how the SCC-based approach can incorporate a number of noteworthy features not shared by the Pigouvian approach, including greater humility. In particular, the Modest Carbon Tax illustrates how an SCC-based approach can be compatible with greater humility about our knowledge of the impacts of climate change, even beyond the methods of uncertainty increasingly central to the SCC estimates themselves. The Modest Carbon Tax also can incorporate greater humility about the limits of governmental regulation—in particular, it is implicitly sensitive to the obvious fact that actual governments are very far from omniscient and omnibenevolent social planners who can be relied upon to perfectly implement perfectly known regulatory "solutions."[6] In particular, the Modest Carbon Tax can implicitly incorporate humility about the limits of governmental knowledge, incentives, and regulatory success, including what levels of mitigation ambition are increasingly likely to trigger large negative unintended consequences, which could at the extreme outweigh the direct good done by a carbon tax. One obvious way the Modest Carbon Tax implicitly incorporates humility about regulation is that it is more likely to be an improvement over the status quo than a Pigouvian tax based on the higher degree of confidence that a lower carbon tax will yield an improvement versus a higher carbon tax equal to the central estimate of the SCC recommended by the Pigouvian approach. Along another dimension, by recommending a simple predictable tax rather than, say, a cap-and-trade system or other mechanisms for putting an implicit or explicit price on carbon, it can avoid some worries about capture and unintended negative consequences that are greater in connection with those other approaches.[7] And finally, by rejecting the welfare maximizing assumption (iii) of the Pigouvian approach, the Modest Carbon Tax becomes consistent with the wide range of reasonable alternative normative views about how to respond to societal challenges like climate change and other social problems. I discuss this last virtue in greater detail in the next section.

Overlapping Consensus, Ecumenical Climate Policy, and the SCC-Based Approach

Ecumenical Climate Policy: How the SCC-Based Approach is Neutral Between Different Reasonable Normative Frameworks

In this subsection I explain why the SCC-based approach can be used in a way that is ecumenical between the wide range of reasonable but incompatible views about the proper goals of government and policy, ranging from mere market-efficiency aimed views, to utilitarian views, to rights-based and other deontological views, to libertarian views, to virtue ethics views, and to other views. Beyond this, in the next subsection, I suggest that the SCC-based approach can help us find an overlapping consensus on a particular climate policy given the range of reasonable but incompatible normative views endorsed in society.

As explained in the previous sections, the most common application of the SCC-based approach is a Pigouvian tax, which aims to minimize the costs of correcting a negative externality, typically with the ultimate policy goal of maximizing benefits via an "optimal" policy (typically assuming other optimal policies are implemented for other market failures). This approach is aimed at market-efficiency and perhaps ultimately a form of utilitarianism/traditional consequentialism.

However, as noted in the previous section, there are other straightforward SCC-based approaches such as the Modest Carbon Tax that do not aim to optimize anything. More generally, the SCC-based approach is compatible with a wide range of normative frameworks, including those that reject traditional consequentialism. Perhaps the easiest way to see this is to consider the alternative normative framework of deontology, which is the classic contrast to utilitarianism/traditional consequentialism. Deontology emphasizes equity, justice, and the avoidance of doing harm as canonical constraints on what actions or policies may be chosen. So, whereas utilitarianism recommends the climate policy that is estimated to have the best wellbeing consequences, deontology might reject that policy because it is unjust, because it lacks sufficient distributional equity, or because it does not do enough to mitigate deaths due to unnecessary emissions by the rich—and instead prefer a different policy that best respects those constraints.

At first glance, it might appear that deontology is incompatible with an SCC-based approach, at least before considering the details of SCC estimates explained above. However, building on the explanation from section "The Social Cost of Carbon (SCC), and Estimates of the SCC" above of the methodology of SCC estimates, we are now in a position to note that in fact SCC estimates aim to quantify deaths as a function of climate change, as well as the locational distribution of both deaths and economic losses.[8] Indeed, SCC estimates are the only existing scientific estimates of death from climate change as a function of different climate policy choices, as well as estimates of the locational distribution of both deaths and economic losses.[9] Thus, from a deontological perspective, these estimates are the natural starting point for evaluating which actions or polices will best avoid causing unnecessary deaths[10] and which policies do best on other criteria of justice and equity.[11]

In fact, these estimates are already used in some applications to evaluate distributional equity (see previous citations) and in the process can be used to educate normative theorists about the difficult and necessary tradeoffs that must be made in climate policy, which normative theorists often ignore. For example, the accounting used to make SCC estimates can make visible the fact that there are deaths "on both sides of the ledger" that much be considered in climate policy, as increased emissions reductions will itself cause increased death via foregone development in poor nations, via higher food prices, etc. So, if one's goal is to evaluate climate policy from a deontological view that aims to prevent causing deaths, then SCC estimates and, in particular, the underlying integrated assessment models and components are the best tool available.[12] (Here I am assuming for ease of exposition that non-consequentialist views have the objective to increase equity-weighted net benefits for society subject to side constraints of the sort relevant to large-scale global policy such as "minimize unnecessary deaths."[13] More nuanced deontological objectives might be preferred by some theorists, but the same point applies that SCC estimates are the best guide available for more nuanced deontological policy recommendation as well).[14]

The discussion so far in this section of how the SCC-based approach is compatible with deontology also generalizes to other reasonable normative theories, given that the core concerns of deontology are one or more of the core concerns of other reasonable normative frameworks for policy analysis (i.e., concerns for justice, equity, and avoiding unnecessary harm). Other important normative frameworks such as virtue ethics, feminist ethics, political liberalism, and global ethics emphasize first and foremost an external critique of the methods of most consequentialist and deontological ethics, and perhaps also emphasize structural features of society as the most urgent locus of normative concern. Their policy recommendations, however, are similarly based on exactly the features that SCC estimates provide the best information about: deaths, economic losses, their socio-economic distribution of deaths and losses, and how structural changes such as carbon pricing and/or climate burden sharing can change these impacts. Note that the SCC-based approach can be used to evaluate optimal policy under the range of leading approaches to distributive justice, including Rawlsian leximin, egalitarianism, and prioritarianism,[15] where these evaluations can estimate the distribution of wellbeing for rich vs. poor in a robust sense that includes a concern for the greater wellbeing impact of dollars lost to the poor than the rich,[16] and can estimate the distribution of non-market changes in death from climate change and air pollution.[17] Thus, the implications of these other normative theories for climate policy are also naturally illuminated by the SCC-based approach and its underlying modeling estimates. In addition, as noted before, the SCC-based approach has the great virtue of further educating these theorists about the difficult and necessary tradeoffs that must be made in climate policy.

Beyond the comprehensive ethical theories discussed so far (which roughly cover the range of reasonable comprehensive theories), the SCC-based approach in conjunction with carbon pricing also has the virtue of respecting individual liberty and the political principles of liberalism. In very brief terms, this is because, first, carbon pricing corrects a market failure that if uncorrected allows freely chosen activities to harm others and make outcomes suboptimal, and, second, because it corrects this market failure in the way that best preserves free choice by simply raising the cost of harmful actions to a level that properly reflects those harms.[18] This is merely an outline of what would need to be a more detailed argument that is outside the focus of this short paper—for more, see the references attached to the preceding endnote, and, for important arguments against this idea, see those attached to this endnote.[19]

Beyond these ethical theories and principles, the SCC-based approach is also ecumenical between the reasonable range of specific "policymaking principles" that might be judged applicable to climate policymaking even in the absence of commitment to any comprehensive normative theory. These principles might be the result of an overlapping consensus or incompletely theorized agreement on how to make policy in the explicit absence of agreement on comprehensive normative theory. For example, consider principles of the form "regulation R should be favored when the aggregate benefits would be X times the aggregate costs and also satisfy distributive justice principle Y." This schema itself describes a wide family of principles resulting from different choices of X and Y, and, for similar reasons noted above, the SCC-based approach and its underlying integrated assessment models provide the best available basis for evaluating the recommendations of these principles for climate policy.

Overlapping Consensus on Climate Policy: How the SCC-Based Approach Can Help Identify a Policy That Is Justifiable to All

Different ethical theories that disagree about how to rank outcomes might nonetheless agree that some policies would be an improvement over the status quo. For example—and continuing the discussion from the end of the previous subsection—incompatible theories might agree for some choice of X, Y, and risk profile that an outcome of new regulation should be preferred over the status quo when aggregate benefits are at least X times greater than costs, the distributive implications at least satisfy minimal equity principle Y, and all of this holds with a sufficient certainty profile. Indeed, the sort of formal work on equity and distributive justice noted in the previous subsection allows for mathematically precise derivations of instances of agreement between specific comprehensive ethical theories.[20] Within the space of agreement, there will often be further agreement about how to rank the outcomes that are agreed to be better than the status quo relative to each other, which implies a frontier of agreement on policy choice between the incompatible theories.

With these conceptual inputs in hand from formal ethics, the SCC-based approach and its underlying estimates are the natural way of adding the climate policy-relevant information that is needed to then calculate the magnitude of mitigation that the different normative theories agree on. If one adds the further assumptions that a carbon tax is the instrument of choice, then the corresponding carbon tax is also thereby calculated to achieve this goal. Notice that the resulting policy recommendation will have the same general structure as the Modest Carbon Tax described above (including the property of being a carbon tax that is some fraction of the Pigouvian tax estimate), except that in this case the tax (or alternative instrument) will be calibrated to reflect the agreement on climate policy between different comprehensive ethical theories in society. In light of this, we might call it the Overlapping Consensus Carbon Tax to differentiate it from the Modest Carbon Tax above (which was simply stipulated to have a particular degree of confidence), while noting that the former will have many of the virtues of the latter—especially for those who are largely on board with the discussion so far but remain skeptical of the assumptions and aspirations of the Pigouvian tax or who at least prefer an approach that prioritizes broad-based political agreement over "Government House Utilitarianism."

The Overlapping Consensus Carbon Tax has some practical advantages given that it is agreeable as an improvement over the status quo to a wider range of political perspectives than, for example, the Pigouvian tax. Thus, identifying the Overlapping Consensus Carbon Tax may have practical political importance, as it may help identify a strategic coordination point for climate policy around which a successful coalition may be built. As one related analogy, the Economists' Statement advocating a carbon tax together with equal per capita refund of the revenues may be seen as an attempt to promote a sufficiently broad-based climate policy to find broad-based support.[21] The fact that such a policy could find a broad consensus might be one reason it has quickly emerged as a focal point of agreement among this diverse group of experts. Empirically, one might see structurally similar political forces (often not normatively justifiable) as generating the actual agreements that explain the less-than-Pigouvian pricing for externalities that we observe in the real world even when externality pricing is the chosen instrument, as in some of the post-1990 Clean Air Act Amendments. The need for consensus provides an explanation and rationale for the more modest-than-Pigouvian pricing approaches that are the norm in real-world market-based environmental policy.

In addition to practical advantages, there are arguably also normative advantages to such an approach; many political theorists argue that political justification requires that structural features of society that are coercively imposed by the government (such as a carbon tax) and have important benefits and burdens for individuals must be supported by an "overlapping consensus" (in the Rawlsian sense) on principles that justify them,[22] or they themselves must be directly supported by an "incompletely theorized agreement" that directly supports those specific policies or principles over alternatives.[23] Thus, an approach like the Overlapping Consensus Carbon Tax might help with difficult normative questions regarding what climate policy is politically justifiable given reasonable pluralism about (principles relevant to) climate policy. It also may have implications for what is justifiable according to some specific comprehensive normative views, especially views such as deontology and/or libertarianism that see the mere promotion of aggregate social welfare as an insufficient justification for coercive policy, and require instead broad-based consent to such coercion as necessary for justification.

Summary of Distinctions Relevant to Evaluating the SCC-Based Approach and Policy Recommendations

The previous sections have argued that the SCC-based approach has more flexibility, can avoid more objections than it may initially appear, and may have additional virtues depending on exactly how one moves from SCC estimates to further conclusions about policy recommendations. In light of the preceding, and taking a step back, we are now in a better position to summarize some of the key distinctions relevant to evaluating the SCC-based approach, as well as to evaluate particular conclusions about policy recommendations based on SCC estimates. I summarize these as a series of questions that yield different evaluations depending on how they are answered:

- Is climate change harmful on balance? I.e., is the true value of the SCC positive?
  - Those who take climate change seriously are committed to answering "yes," in which case the answer to this question gives no reason for skepticism about the SCC-based approach. (Furthermore, even if it were true, contrary to general academic consensus, that the global net costs of carbon were negative (e.g. if the net benefits from warming were sufficiently large to exceed the large costs), the distribution of costs and benefits would still matter and be highly unequal, and thus GHG emitters would still be imposing net costs on many others (including imposing almost all climate costs on third parties); this would generate equity-based reason (including from the perspective of liberalism or even principled libertarianism) for carbon pricing to help address this disparity).[24]
- Are SCC models the best tool for making estimates of the SCC, including the range of uncertainty over the true value of the SCC?
  - Mainstream climate economics and governmental cost–benefit studies argue that the answer is "yes," but some experts have disagreed even while taking climate change seriously (see Pindyk,

for example). I set aside this issue in this paper, except to note that even critics of SCC models such as Pindyk often believe that the SCC should still play a primary role in climate policy, albeit with estimates using different methods, such as expert elicitation.[25]

- Should the magnitude of our emissions reduction ambition simply reflect the central estimate of the SCC, as in the Pigouvian tax approach? Or if not, what are the further factors that are relevant to deciding what policy implementation we should favor and disfavor? For example, do we need to further incorporate worries about the costs of policy, feasibility, political economy, justification, unforeseen negative side effects of governmental controls, ethical reasons that climate policy should explicitly reflect weights for equity and justice, etc.? Should our ambition be calibrated to an overlapping consensus between incompatible normative theories that represent the range of reasonable views in society?
  - Many argue that there are further relevant factors of equity, political justifiability, and the like not accounted for by the Pigouvian tax approach, even assuming correct estimates of the SCC.[26] The Modest Carbon Tax (and the discussion in the last two sections) illustrates how endorsing the importance of these further factors is nonetheless consistent with endorsing the SCC-based approach.
- Should carbon pricing be one main aspect of our overall response to climate change?
  - Most economists and the vast majority of climate policy experts argue that the answer is "yes" (even while some also argue that other policies or regulations are needed as well[27]), which provides natural motivation for the SCC-based approach to the carbon pricing component of an overall response to climate change.[28] This is for many reasons, including greater efficiency, creating the best incentives for green innovation,[29] and less legal and political economy risks.[30] In addition, even if one answered "no," SCC estimates might still be useful in deciding the magnitude of emission reductions to aim for via other policy instruments. Beyond noting these arguments, I set aside this issue in this paper.

- Should a carbon tax be the main policy instrument for carbon pricing or something else?
  - A majority of economists and climate policy experts argue that the answer is "yes" and thus that a carbon tax is preferable to other forms of carbon pricing, such as a global cap-and-trade system. This is largely due to transparency, reduced worries about corruption, and the advantages of a more predictable price signal than with alternative forms of carbon pricing.[31] Thus, a carbon tax has many advantages from the perspective of those who stress modesty in climate policy proposals. Beyond noting these advantages, I set aside this issue in this paper.

In sum, endorsing the SCC-based approach as the best way forward for climate policy is consistent with endorsing many worries about mainstream climate policy recommendations and is consistent with skepticism about many possible applications of an SCC-based approach, such as the Pigouvian approach.

Additional Worries About the SCC-Based Approach

There are many important residual worries about the SCC-based approach that might be developed in more detail. In this section, I'll highlight a few important worries that are especially prominent in contemporary discourse and policy debates, and I'll also provide a quick outline of some possible replies, without pretending to settle the issues raised. Most of these worries fall under the partially overlapping headings of political economy, uncertainties in (or unrealism of) SCC estimates, and political feasibility. I suggest that even if we agree that all of these objections are on the right track or at least have a kernel of truth in their criticism of existing SCC-based recommendations, nonetheless the correct response to climate change still involves substantial emissions reductions to be achieved via policy, and the best methods for deciding the magnitude of those reductions still depend essentially on next generation versions of the SCC-based approach.

Worry: Too Much Mitigation, Because of Overly Pessimistic Assumptions About Technology and Adaptation

The first worry is that SCC-based analyses recommend too high of emissions reductions via too high of a carbon price (or other government controls), because SCC estimates ignore the possibility of exogenous technological solutions to climate change and more generally make highly uncertain and pessimistic assumptions about many other factors including endogenous technological change, adaptation, and many other things.[32]

There are many important dimensions to this worry. One dimension of the worry is simply that the GHG reductions that we should aim for with policy are actually less than SCC models tend to estimate, because the models ignore or have overly pessimistic assumptions about the propensity of technology to endogenously respond to problems when incentives and desire exist to do so (e.g. quickly innovate with new technology that reduces the carbon intensity of the economy more quickly than models assume and/or innovate with new technology to remove GHGs from the atmosphere).

In addition, a complementary worry is that the models do not comprehend exogenous technology changes that might solve the climate problem to some large extent in a way that is independent of climate policy-related incentives or preferences. For an oft-cited example, in the early 1900s there was great concern about the levels of horse manure accumulating on the streets of major cities. There was a lot of discussion about the correct policy to address horse manure, and there were many suggestions that radical intervention and policy controls were needed. But in fact the problem was solved entirely by exogenous technology changes—in other words changes that had nothing to do with any policy having to do with horse manure or any related effort to specifically address the horse manure problem. The solution was the unrelated and independent invention of the automobile and the decline of horse-driven transportation. Similarly, one might think that the arrival of e.g., cold fusion based on research that is not driven by climate policy or, perhaps more realistically, some combination of technology that we haven't even imagined yet and that it will arise due to purely entrepreneurial and creative processes that have nothing to do with climate policy might largely deal with the climate problem effectively themselves.[33] The social cost of the carbon models do not comprehend these exogenous technology changes.

The preceding summarizes a number of ways the worry might arise that SCC models recommend too much government control based on overly pessimistic assumptions about technology and adaptation. If taken further and in a perhaps more libertarian direction, the worry could even be expressed that a Pigouvian tax as recommended by SCC estimates could even be net harmful to welfare versus something more in the direction of a more free market approach.

To begin to articulate a response to this worry, we can return to the preliminary observation from an earlier section that the true value of the social cost of carbon is positive, which everyone who takes climate change seriously is committed to. As a further a priori matter, we can note that as we move from zero to increasingly positive numbers for the SCC, every individual's confidence that the SCC is at least as high as the number in question will decrease. For any individual evaluating climate policy, there should thus be a positive SCC of quantity N such that the individual assigns very high confidence to the SCC being at least that value, as shown in Fig. 1.

With this in mind, we can see more clearly that much of the force of this worry is really about relying on central estimates of the SCC (as the Pigouvian approach does) rather than a more modest approach—and if we focus instead on a more modest carbon tax, much of the force of this worry dissipates. In other words, the important kernel of truth in this worry is really about over-reliance on central estimates from the SCC models—especially in connection with deterministic models from previous decades, which were based on particular estimates and structural assumptions of those specific instances of SCC models. SCC modelers acknowledge these worries and are already updating estimates to better explore uncertainty about the assumptions about technology, adaptation, and beyond (along with other important uncertain components, such as [especially] climate damages, climate sensitivity, and the climate system).[34]

These observations about the flexibility of the SCC approach and recent improvements in estimates are not a fully decisive reply to the worry in this section, which is that there is still too much pessimism even within the newest generation of SCC estimates. However, when these observations are combined with the arguments in sections above, including the possibility of a more modest SCC-based approach such as the Modest Carbon Tax, there are arguably straightforward ways for policy recommendations to take this worry on board, and there remain good reasons to endorse the SCC-based approach as at least one centrally important guide to climate policy.

Worry: Too Much Mitigation, Given Political Infeasibility and Counterproductive Blowback of Recommended Mitigation

With the preceding discussion in hand, the next important worry to consider is that the SCC-based approach recommends too high of emissions reductions via too high of a carbon price, because high carbon prices recommended by that approach would not be politically feasible and would lead to counterproductive blowback if implemented. As one real-world source of worry along this line, many would cite our experience with the Yellow Vests Movement and other protests in response to carbon pricing and related policies.

This is an important worry. One thing to note is that this is a worry not about the correctness of SCC estimates, but rather this is a worry about what policy should actually be implemented given the SCC, taking into account feasibility, political economy, and beyond. When this observation is combined with the arguments in sections above, including the possibility of a more modest SCC-based approach, arguably there are straightforward ways to revise policy recommendations in light of this worry (including simply moving in the direction of a more modest tax), and thus good reasons remain to endorse the SCC-based approach as at least one centrally important guide to climate policy.

Worry: Too Much Mitigation, Because Domestic SCC Is Only a Fraction of Global SCC

The next important worry to consider is that SCC-based analyses recommend too high of emissions reductions via too high of a carbon price, because they ignore the global collective action problem if they are interpreted as offering recommendations of what policies national governments should adopt.[35]

This is an important worry, especially as it is uncontroversial that, for example, the USA share of global climate damages is only somewhere on the order of 10–15% of global damages.[36] As a direct consequence, some commentators have argued that the correct carbon price to implement should only be 10–15% of the (global) SCC, since this would be the amount that internalizes the damages to USA citizens. Note again that this is a worry not about SCC estimates themselves, but rather about what policy should actually be implemented given these estimates, taking into account feasibility, political economy, and, in this case, the legitimate aims of domestic policymaking.[37]

At the same time, pushing back on this worry a little, even if we agree that there is an important kernel of truth here, it is important to emphasize that nations like the USA have self-interested reasons to spur cooperation on climate, and therefore they presumably have self-interested reasons to implement some climate action to spur global cooperative action. In addition, widely endorsed policy instruments such as border tax adjustments can mitigate this worry,[38] and the domestic use of revenue from a carbon tax[39] and decreased domestic air pollution[40] can generate large benefits domestically. And of course there are important normative arguments for focusing on the global social costs of carbon emissions as well as domestic costs.[41]

In sum, much like the previous points, some reason may emerge for favoring a more modest carbon tax than a Pigouvian tax based on the sum of the previous three worries. However, as we've seen there are straightforward ways to revise policy recommendations in light of this worry within the framework of the SCC approach (including simply moving in the direction of a more modest tax), and there remain other good reasons to endorse the SCC-based approach as at least one centrally important guide to climate policy.

Worry: Too Little Mitigation, Because Worst-Case Scenarios Are Ignored and/or There Is Too Much Pure Time Preference (i.e., Impacts More Than a Century from Now Are Given Little Importance)

The next important worry to consider is that SCC-based analyses recommend too low of emissions reductions via too low of a carbon price, because SCC estimates ignore worst-case scenarios by focusing only on median parameter estimates or low-tail risk distributions across uncertain parameters, or they have too much pure time discounting.[42] This is an important worry, and we should be particularly mindful to acknowledge shortcomings of SCC estimates "on both sides," such as this worry that the SCC should be higher (in contrast to previous worries that in one sense or other put pressure in the other direction).

Unlike previous worries, this worry is really confined to specific parameter choices in SCC calculations, or specific deterministic model structures (rather than those that more adequately represent uncertainty, the range of possible outcomes, and expected value). In response, SCC estimates can be updated to whatever choice of parameters is desired, and next-generation models also aim to natively incorporate uncertainty. Thus, there are straightforward ways to revise policy recommendations in light of this worry, and, in fact, the most natural revisions assume the SCC framework for their implementation, so no reason emerges here to reject the SCC-based approach as at least one centrally important guide to climate policy.

Worry: Too Little Mitigation, Because Many Harms Are Not Accounted for, Especially Harm to the Poor, Oppressed, and Most Vulnerable

The next important worry to consider is that SCC-based analyses recommend too low of emissions reductions via too low of a carbon price, because SCC estimates ignore many harms from climate change,[43] as well as inequalities within societies and the greater harm to the poor within societies from climate change,[44] including ignoring the possibilities for use of revenues from carbon taxes to alleviate poverty and inequality.[45] More controversially, it could be argued that SCC estimates ignore important harms to non-human animals and ecosystems.[46] This is an important worry, arguably correct when applied to specific SCC estimates of the past (see previous citations)—and it again highlights that we should be particularly careful to acknowledge shortcomings of SCC estimates "on both sides," such as this worry that the SCC should be higher (in contrast to previous worries in initial sections that in one sense or other put pressure in the other direction).

As with the previous worry, "internal" corrections within the SCC framework are the most natural way of taking this worry properly into account, which generates revised SCC estimates when (e.g.) "damage functions" are adjusted upward that describe the likely magnitude of the harms as a function of climate change. Exactly this kind of update is currently in progress for the main SCC estimates used in US government analyses, guided by recommendations from a recent National Academies report.[47] Again, no special reason emerges here to resist the SCC-based approach as at least one centrally important guide to climate policy.

Worry: Carbon Taxation Is the Wrong Mechanism for Emissions Reductions

Finally, a different important worry to consider is that carbon taxation and other policy controls recommended by SCC-based policy evaluations end up recommending the wrong mechanisms for emissions reductions, because (this worry claims) of the political infeasibility of carbon pricing as the sole mechanism for climate policy. Thus, according to some proponents of this, the superiority in some specific applications of command-and-control policies and governmental investment in subsidies and technology.[48]

This is an important worry and, like Worry 2, is a worry not about SCC estimates themselves, but rather about how they should be translated into policy, especially what policy instruments should be chosen to achieve emissions reductions. Furthermore, this issue does not immediately tell in the direction of more or less ambition, unlike the worries above.

However, if the upshot of this worry is that carbon pricing is needed but must be supplemented with large other policies, then it is possible that the optimal portfolio of policies might include (e.g.) a more modest carbon tax than a purely Pigouvian tax-based climate policy would recommend.

In sum, the preceding worries and responses indicate that even if we agree that all of these objections above to existing SCC-based analyses have an important kernel of truth, they do not undermine arguments that the correct response to climate change still involves substantial emissions reductions to be achieved via policy, and the best methods for deciding the magnitude of those reductions still depend essentially on the SCC-based approach. Indeed, many of these objections pull in opposite directions (i.e. in the direction of higher vs. lower estimates) and thus may in some cases offset one another to some important extent. The upshot is that even those who agree with these important worries do not thereby have good reason for wholesale rejection of policy guided by SCC estimates, and in fact they should still find action-guiding value in those estimates as one of the best available tools for deciding what we should do about climate change.

Acknowledgement Thanks to Jonathan Adler, Karen Bradshaw, Frank Errickson, Alison Grant, Mark Pennington, Danny Shahar, John Thrasher, Greg Wolcott, Cayla Xue, and participants in the Institute for Humane Studies seminar on liberty, property, and pollution; special thanks to Jonathan Adler and Greg Wolcott for their feedback and leadership.

Notes

1. William P. Nordhaus and Paul Sztorc, DICE 2013R: Introduction and User's Manual (2013), online at https://aida.econ.yale.edu/~nordhaus/homepage/documents/DICE_Manual_103113r2_000.pdf. Accessed October 11, 2017.
2. Arthur Cecil Pigou, *The Economics of Welfare* (London: Macmillan & Co., 1920); William Nordhaus, "How Fast Should We Graze the Global Commons?" *The American Economic Review* 72, no. 2 (May 1982): 242–246; Kevin Rennert, Brian Prest, William Pizer, Richard G. Newell, David Anthoff, Cora Kingdon, Lisa Rennels, Roger Cook, Adrian E. Raftery, Hana Ševčíková, and Frank Errickson, "The Social Cost of Carbon: Advances in Long-Term Probabilistic Projections of Population, GDP, Emissions, and Discount Rates," *Brookings Papers on Economic Activity*, September 9, 2021.
3. Jay K. Rosengard and Joseph E. Stiglitz, *Economics of the Public Sector*, 4th ed. (New York, NY: W. W. Norton & Company, 2015); *see also* Thomas H. Tietenberg, *Environmental and Natural Resource Economics*, 11th ed. (Oxfordshire: Routledge, 2018).
4. National Academies, *Valuing Climate Damages: Updating Estimation of the Social Cost of Carbon Dioxide* (Washington, DC: The National Academies Press, 2017); William D. Nordhaus, "Revisiting the Social Cost of Carbon." *Proceedings of the National Academy of Sciences* 114, no. 7 (January 31, 2017): 1518–1523. https://doi.org/10.1073/pnas.1609244114; Rennert et al., "Social Cost."
5. 95% confidence is arbitrarily chosen here, largely for purposes of a clear example and clear alternative with different properties than the Pigouvian tax approach. Who exactly should be in charge of estimating this confidence interval? It currently is the job of an interagency working group within the federal government, which in the future could be put under the administrative control of a more formal bipartisan institution. Note also that our confidence in the likely outcomes of policy might increase with experience, leading to ratcheting up of policy conditional on the success of a modest carbon tax.
6. Friedrich Hayek, "The Use of Knowledge in Society," *The American Economic Review* 35, no. 4 (1945): 519–530.
7. Tietenberg, *Economics*; see also Jonathan H. Adler, "The Legal and Administrative Risks of Climate Regulation," *Environmental Law Reporter* 51 (2021): 10485.

8. Some SCC models are more 'reduced form' in the sense that they only implicitly represent deaths, and in some cases do not explicitly represent the spatial distribution of damages. But many SCC models and model components are explicit in their representation of all of these things, i.e., deaths, economic losses, and their spatial distribution. See Delavane Diaz and Frances Moore, "Quantifying the Economic Risks of Climate Change," *Nature Climate Change* 7, no. 11 (November 2017): 774–782; *see also* National Academies, *Valuing Climate Damage*.
9. Delavane Diaz and Frances Moore, "Quantifying the Economic Risks of Climate Change," *Nature Climate Change* 7, no. 11 (November 2, 2017): 774–782. https://doi.org/10.1038/nclimate3411; Rennert et al., Social Cost of Carbon; Tamma Carleton et al., "Valuing the Global Mortality Consequences of Climate Change Accounting for Adaptation Costs and Benefits," *The Quarterly Journal of Economics*, April 21, 2022. https://doi.org/10.1093/qje/qjac020.
10. Josh Broome, "How Much Harm Does Each of Us Do?" in *Philosophy and Climate Change*, eds. Mark Budolfson, Tristram McPherson, and David Plunkett (Oxford: Oxford University Press, 2021), 281–291.
11. Francis Dennig, Mark Budolfson, Marc Fleurbaey, Asher Siebert, and Robert H. Socolow, "Inequality, Climate Impacts on the Future Poor, and Carbon Prices," *Proceedings of the National Academy of Sciences* 112, no. 52 (December 7, 2015): 15827–15832, https://doi.org/10.1073/pnas.1513967112; Mark Budolfson, David Anthoff, Francis Dennig, Frank Errickson, Kevin Kuruc, Dean Spears, and Navroz K. Dubash, "Utilitarian Benchmarks for Emissions and Pledges Promote Equity, Climate and Development," *Nature Climate Change* 11, no. 10 (September 13, 2021): 827–833, https://doi.org/10.1038/s41558-021-01130-6; Mark Budolfson, Francis Dennig, Frank Errickson, Simon Feindt, Maddalena Ferranna, Marc Fleurbaey, David Klenert, et al., "Climate Action with Revenue Recycling Has Benefits for Poverty, Inequality and Well-Being," *Nature Climate Change* 11, no. 12 (November 29, 2021): 1111–1116, https://doi.org/10.1038/s41558-021-01217-0; Frank Errickson, Klaus Keller, William D. Collins, Vivek Srikrishnan, and David Anthoff, "Equity Is More Important for the Social Cost of Methane Than Climate Uncertainty," *Nature* 592, no. 7855 (April 21, 2021): 564–570, https://doi.org/10.1038/s41586-021-03386-6.
12. Broome, "How Much."

13. Further complications arise if one considers deontological theories that cannot be easily represented as maximizing subject to side constraints—for more discussion see the literature on consequentializing deontological theories. See Paul Hurley, "Consequentializing," in *Oxford Handbook of Consequentialism*, ed. Douglas Portmore (Oxford: Oxford University Press, 2021), 24–35.
14. Climate Analytics and NewClimate Institute, The Climate Action Tracker, 2009, https://climateactiontracker.org/; Civil Society Equity Review Group, "Fair Shares: A Civil Society Equity Review of INDCS," *Human Rights Documents Online*, February 22, 2018, https://doi.org/10.6084/m9.figshare.5917399.v2.
15. Derek Parfit, *Equality or Priority? The Lindley Lecture* (University of Kansas, 1995); Marc Fleurbaey, "Equality versus Priority: How Relevant Is the Distinction?," *Economics and Philosophy* 31 (2015): 203–217; Matthew D. Adler, David Anthoff, Valentina Bosetti, Greg Garner, Klaus Keller, and Nicolas Treich, "Priority for the Worse-off and the Social Cost of Carbon," *Nature Climate Change* 7, no. 6 (2017): 443–449, https://doi.org/10.1038/nclimate3298; Matthew D. Adler, *Measuring Social Welfare: An Introduction* (Oxford: Oxford University Press, 2019).
16. Dennig et al., "Inequality"; Budolfson et al., "Utilitarian Benchmarks."
17. Noah Scovronick, Mark Budolfson, Francis Dennig, Frank Errickson, Marc Fleurbaey, Wei Peng, Robert Socolow, Dean Spears, and Fabian Wagner, "The Impact of Human Health Co-benefits on Evaluations of Global Climate Policy," *Nature Communications* 10, no. 2095 (2019): 1–12; Tamma Carleton, Amir Jina, Michael Delgado, Michael Greenstone, Trevor Houser, Solomon Hsiang, Andrew Hultgren, et al., "Valuing the Global Mortality Consequences of Climate Change Accounting for Adaptation Costs and Benefits," *The Quarterly Journal of Economics* (April 21, 2022), https://doi.org/10.1093/qje/qjac020.
18. See Mark Budolfson, "Market Failure, the Tragedy of the Commons, and Default Libertarianism in Contemporary Economics and Policy," in *Oxford Handbook of Freedom*, eds. David Schmidtz and Carmen Pavel (New York: Oxford University Press, 2017), 257–282; Mark Budolfson, "Arguments for Well-Regulated Capitalism, and Implications for Global Ethics, Food, Environment, Climate Change, and Beyond," *Ethics & International Affairs* 35, no. 1 (May 7, 2021): 83–98, https://doi.org/10.1017/s0892679421000083. Indeed, arguably for anything like economic liberalism to function adequately or in an ethically defensible

manner, something akin to a carbon price might be morally mandatory. Ibid.

19. See Mark Pennington, Climate Change, Political Economy, and the Problem of Comparative Institutions Analysis this volume; Dale Jamieson and Marcello Di Paola, "Climate Change, Liberalism, and the Public/Private Distinction," in *Philosophy and Climate Change*, eds. Mark Budolfson, Tristram McPherson, and David Plunkett (Oxford: Oxford University Press, 2021), 370–396; https://doi.org/10.1093/oso/9780198796282.003.0017.
20. Matthew D. Adler and Marc Fleurbaey, *The Oxford Handbook of Well-Being and Public Policy* (Oxford: Oxford University Press, 2016); Gustaf Arrhenius, Mark Budolfson, and Dean Spears, "Does Climate Change Policy Depend Importantly on Population Ethics? Deflationary Responses to the Challenges of Population Ethics for Public Policy," in *Philosophy and Climate Change*, eds. Mark Budolfson, Tristram McPherson, and David Plunkett (Oxford: Oxford University Press, 2021), 111–136; Bernardo A. Bastien-Olvera and Frances C. Moore, "Use and Non-Use Value of Nature and the Social Cost of Carbon," *Nature Sustainability* 4, no. 2 (September 28, 2020): 101–108, https://doi.org/10.1038/s41893-020-00615-0.
21. Climate Leadership Council, "Economists' Statement," *The Wall Street Journal*, January 17, 2019, https://clcouncil.org/economists-statement/.
22. An overlapping consensus in the Rawlsian sense is based on normative reasons, including reasons to find and respect agreement that reflects proper normative and empirical humility. This distinguishes an overlapping consensus from mere strategic/pragmatic agreements, which are potentially unstable and lack normative justification. See John Rawls, *Political Liberalism* (New York, NY: Columbia University Press, 1993).
23. See Cass R. Sunstein, "Incompletely Theorized Agreements," *Harvard Law Review* 108, no. 7 (1995): 1733–1772.
24. See Jonathan H. Adler, "Taking Property Rights Seriously: The Case of Climate Change," *Social Philosophy & Policy* 26, no. 2 (2009).
25. Robert Pindyck, "The Use and Misuse of Models for Climate Policy," *Review of Environmental Economics and Policy* 11 (2015): 100–114, https://doi.org/10.3386/w21097.
26. Henry Shue, "Subsistence Emissions and Luxury Emissions," *Law & Policy* 15, no. 1 (January 1993): 39–60, https://doi.org/10.1111/j.1467-9930.1993.tb00093.x; David Anthoff, Cameron Hepburn, and Richard S.J. Tol, "Equity Weighting and the Marginal Damage Costs of Climate

Change," *Ecological Economics* 68, no. 3 (2009): 836–849, https://doi.org/10.1016/j.ecolecon.2008.06.017; Stephen M. Gardiner, *Climate Ethics: Essential Readings* (Oxford: Oxford University Press, 2010); Nicholas Stern, *Why Are We Waiting?:The Logic, Urgency, and Promise of Tackling Climate Change* (Cambridge, MA: The MIT Press, 2015); Frank C. Errickson, Klaus Keller, William D. Collins, Vivek Srikrishnan, and David Anthoff, "Equity Is More Important for the Social Cost of Methane Than Climate Uncertainty," *Nature* 592, no. 7855 (April 21, 2021): 564–570, https://doi.org/10.1038/s41586-021-03386-6; Budolfson et al., "Utilitarian Benchmarks"; Geofrrey Heal, "Economic Aspects of the Energy Transition," *Environmental and Resource Economics* (March 7, 2022), https://doi.org/10.1007/s10640-022-00647-4.

27. For example, almost all economists agree that carbon pricing is necessary in order to create needed economy-wide incentives to cause rational decarbonization across all aspects of society, which leaves open the possibility (as some have argued) that other complementary policies are necessary as well (e.g., investments in hydrogen, or direct air capture, etc.). See Heal, "Economic Aspects"; Stern, *Why Are We Waiting?*
28. Climate Leadership Council, "Economists' Statement."
29. Tietenberg, *Environmental and Natural Resource Economics.*
30. Adler, "Legal and Administrative Risks."
31. Climate Leadership Council, "Economists' Statement"; *see also* Jonathan Adler, "The Legal and Administrative Risks of Climate Regulation."
32. Matt Ridley, *The Rational Optimist: How Prosperity Evolves* (New York, NY: Harper Perennial, 2011); Pennington, "Climate Change, Political Economy, and the Problem of Comparative Institutions Analysis."
33. Steven D. Levitt and Stephen J. Dubner, *Freakonomics: A Rogue Economist Explores the Hidden Side of Everything* (New York, NY: William Morrow, 2006).
34. National Academies, "Valuing Climate Damages."
35. Art Fraas, Randall Lutter, Susan Dudley, Ted Gayer, John Graham, Jason F. Shogren, and W. Kip Viscusi, "Social Cost of Carbon: Domestic Duty," *Science* 351, no. 6273 (2016): 569.
36. Nordhaus and Sztorc, DICE.
37. Note that this intersects with philosophical debates about statism vs. cosmopolitanism, and the like. See John Rawls, *The Law of Peoples: With "The Idea of Public Reason Revisited"* (Cambridge, MA: Harvard University Press, 1999); Charles Beitz, *Political Theory and International*

Relations, rev. ed. (Princeton: Princeton University Press, 1999); Laura Valentini, *Justice in a Globalized World* (New York: Oxford University Press, 2011).
38. Climate Leadership Council, "Economists' Statement."
39. Climate Leadership Council, "Economists' Statement"; Budolfson et al., "Climate Action with Revenue Recycling."
40. Scovronick, et al., "Impact of Human Health Co-benefits."
41. Celine Guivarch Aurèlie Mèjean, Antonin Pottier, and Marc Fleurbaey, "Social Cost of Carbon: Global duty," *Science* 351, no. 6278 (2016): 1160–1161; Kian Mintz-Woo, "Two Moral Arguments for a Global Social Cost of Carbon," *Ethics, Policy & Environment* 21, no. 1 (2018): 60–63; Paul Kelleher, *The Social Cost of Carbon: Ethics and the Limits of Climate Change Economics*, forthcoming.
42. Nicholas Stern, *The Economics of Climate Change*, 1st ed. (Cambridge: Cambridge University Press, 2007); Pindyck "The Use and Misuse of Models"; Gernot Wagner and Martin L. Weitzman, *Climate Shock: The Economic Consequences of a Hotter Planet*, 1st ed. (Princeton, NJ: Princeton University Press, 2016).
43. National Academies, "Valuing Climate Damages."
44. Dennig, et al., "Inequality."
45. Thomas Sterner, *Fuel Taxes and the Poor: The Distributional Effects of Gasoline Taxation and Their Implications for Climate Policy*, 1st ed. (Washington, DC: RFF Press, 2012); Dennig et al., "Inequality"; Budolfson et al., "Climate Action with Revenue Recycling."
46. Wayne Hsiung and Cass R. Sunstein, "Climate Change and Animals," *University of Pennsylvania Law Review* 155, no. 6 (2007): 1695–1740, https://scholarship.law.upenn.edu/penn_law_review/vol155/iss6/7/; *see also* Mark Budolfson and Dean Spears, "Optimal Climate Policy including Animal Welfare" (Princeton CFI Working Paper 2018); Bernard A. Bastien-Olvera and Frances C. Moore, "Use and Non-Use Value of Nature and the Social Cost of Carbon," *Nature Sustainability* 4, no. 2 (September 28, 2020): 101–108, https://doi.org/10.1038/s41893-020-00615-0.
47. National Academies, "Valuing Climate Damages."
48. Stern, *Why Are We Waiting?*; Heal, "Economic Aspects of the Energy Transition."

Index

J. H. Adler (ed.), *Climate Liberalism*, Palgrave Studies in Classical Liberalism,
https://doi.org/10.1007/978-3-031-21108-9

The manufacturer's authorised representative in the EU is Springer Nature Customer Service Centre GmbH, Europaplatz 3, 69115 Heidelberg, Germany. If you have any concerns regarding our products, please contact ProductSafety@springernature.com

Printed and bound by CPI Group (UK) Ltd, Croydon, CR0 4YY
30/09/2026
02225243-0015